Édouard Hospitalier, Gordon Wigan

The electrician's pocket-book

The English edition of Hospitalier's

Édouard Hospitalier, Gordon Wigan

The electrician's pocket-book
The English edition of Hospitalier's

ISBN/EAN: 9783337228996

Printed in Europe, USA, Canada, Australia, Japan

Cover: Foto ©berggeist007 / pixelio.de

More available books at **www.hansebooks.com**

THE ELECTRICIAN'S POCKET-BOOK

THE ENGLISH EDITION OF

HOSPITALIER'S

"Formulaire Pratique de l'Electricien;"

TRANSLATED, WITH ADDITIONS,

BY

GORDON WIGAN, M.A.,

BARRISTER-AT-LAW, MEMBER OF THE SOCIETY OF TELEGRAPH ENGINEERS
AND ELECTRICIANS.

CASSELL & COMPANY, LIMITED:
LONDON, PARIS & NEW YORK.
1884.

[ALL RIGHTS RESERVED.]

AUTHOR'S PREFACE
TO THE SECOND YEAR'S EDITION.

The favourable reception accorded to our FORMULAIRE, and the advice and encouragement which we have received, prove that we were not mistaken as to the purpose and utility of this little book. We therefore now bring it forward in the second year, with more confidence than in the first year of its issue. Each one of the numerous alterations to be found in this edition is an improvement, as it either fills up a gap, corrects an error, or gives some new information.

Thus, in the first part, which is devoted to first principles, definitions, and general laws, we have remodelled the whole of the part on induction, and given formulæ for the galvanic field produced by a current of any given common geometrical form.

We have also modified some of the definitions of the units of measurement, in order to make them clearer and more precise; we have given the "Thomson's bridge" method for the measurement of very small resistances, the formulæ of Thomson's new voltmeters and ammeters, the formulæ of the bifilar suspension, and the work produced by men and horses, etc.

In the Fourth Part, devoted to applications of Electricity, although but few new inventions have appeared, we have, nevertheless, been enabled to add the results of tests of the new batteries of Skrivanow, Lalande, and Chaperon; the Edison-Hopkinson, Schuckert, and Ferranti dynamos; Ayrton and Perry's motors; and the Grenoble experiments of Marcel Deprez on electrical transmission of power.

We have completed the chapter on alternating current machines by giving the methods of Joubert and Potier, which enable the electromotive force, current strength, and energy of such machines to be measured. The methods are fully given, so that either the efficiency of transformers or secondary generators, which are now receiving so much attention, or the high electromotive forces used in the transmission of energy, may be easily measured.

In conclusion, we beg to thank our correspondents for their valuable co-operation. They will see that, as far as possible, we have taken advantage of their advice and information.

We trust that they will kindly continue their friendly collaboration, and again, in the common interest, help us still further to improve this volume.
E. H.

Paris, February, 1884.

TRANSLATOR'S PREFACE.

In common with many of those who have had to do with the modern development of electrical engineering, I had long desired some small portable book, in which it would be easy to find constants, formulæ, methods, and other practical information in a concise form and without difficulty. The favourable reception accorded to M. Hospitalier's "Formulaire Pratique de l'Électricien" by English electricians, has led me to hope that a similar book in English might be useful to electrical engineers.

I have therefore prepared the present work, which consists principally of a translation of M. Hospitalier's "Formulaire." This, the main portion of it, has been carefully compared with M. Hospitalier's edition for 1884, and almost all the additional information contained in that edition has been added.

During the progress of the work I have also added such new matter as my own reading or experience has suggested.

My thanks are due to Prof. Fleeming Jenkin, F.R.S., and other friends, for valuable suggestions; to Mr. W. H. Preece, F.R.S., and Mr. Kempe, of the Post Office, who have kindly allowed me to make use of some of the technical instructions of their department; and also to Profs. Ayrton and Perry, and Messrs. Crompton and Kapp, for most valuable contributions.

I trust, that though I may not have added very much in quantity to the information to be found in the French edition, the additions may yet be found of value, and that at all events I may have succeeded in producing a faithful and intelligible rendering of the original.

It is intended, should the work meet with a favourable reception, to follow M. Hospitalier's example and republish it periodically, with such additions as may be desirable. I therefore venture to repeat his appeal, and to ask electricians generally to assist in such a task by forwarding to me, addressed to the publishers, corrections of any errors which may be found in this book, suggestions for its improvement, and, above all, the results of any new work, in the form of tables or formulæ.

In conclusion, I must express my great indebtedness to Mr. Alfred J. Frost, Librarian of the Society of Telegraph Engineers and Electricians, who, at a time when I was unable to attend to any business, undertook the somewhat formidable task of correcting the final proofs and seeing this work through the press.

G. W.

TABLE OF CONTENTS.

(*See also Index to Tables, page* 315.)

FIRST PART.

Definitions—First Principles—General Laws.

	PAGE
Magnetism.—Definitions—Laws of magnetic actions—Properties of lines of force—Coefficient of induced magnetism—Magnetic induction—Terrestrial magnetism	1
Static electricity.—Law of electrical attraction and repulsion—Distribution of electrostatic charges—Induction—Specific inductive capacity	5
Condensers.—Capacity—Charge—Condensers arranged parallel and in cascade—Energy in a condenser—Contact electricity—Volta's law	7
Dynamic electricity, or electricity in movement. Laws of currents.—Ohm's law—Kirchoff's laws—Bosscha's corollaries—Specific resistance—Conductivity—Resistance of conductors—Resistance of derived shunt or parallel circuits	9
Voltaic batteries.—Laws of chemical actions—Constants—Energy of a battery—Arrangement of cells—Pollard's theorem	11
Electrolysis.—Faraday's laws	14
Heating effect of currents.—Joule's law	15
Work produced by currents	16
Electro-dynamics.—Electric or galvanic field—Magnetic shell—Solenoid—Mutual actions of two currents—Rectilinear currents—Plane rectangular circuits—Action of two coils at a distance—Ampère's law—Action of the earth on currents—Astatic conductors—Solenoids	18
Electro-magnetism.—Fundamental principle—Ampère's rules—Multiplier—Action of currents on magnets—Electro-magnet—Maxwell's rule—Action of an element of a current on a magnet	

pole—Galvanic field produced by an arc of a circle, a circle, an infinite rectilinear circuit, and a plane closed circuit—Magnetisation by currents—Rule for finding the poles of electro-magnets .	20
Induction.—Laws of induction—Extra current—Induction in a rectilinear circuit displaced so as to be parallel to itself in a uniform magnetic field—Induction in a closed circuit—Influence of the extra current on induced currents—Lenz's law—Conservation of energy in induction	24

SECOND PART.

Units of Measurement.

Fundamental units.—C.G.S. system—Multiples and sub-multiples—Decimal notation—Dimensions—Derived units . . .	29
Geometrical units.—Length, area, volume — Units of different countries	31
Mechanical units.—Velocity—Acceleration—Units of force and weight—Units of work or energy—Horse-power—Watt . .	35
Magnetic units.—Unit pole—Unit magnetic field . . .	40
Electro-magnetic units.—C.G.S. units—Practical units—Units of current strength, quantity, electromotive force, resistance, and capacity	41
Comparison of electrical units used by different physicists . .	43
Units used by the house of Siemens at Berlin	47
Electrostatic units	47
Various units.—Pressure—Temperature—Heat — Mechanical equivalent of heat—Units of energy—Photometric units . . .	48

THIRD PART.

Measuring Instruments and Methods of Measurement.

Geometrical measurements.—Micrometer gauge	53
Mechanical measurements.—Velocity, force, and work . .	54

ELECTRICAL MEASUREMENTS.

Resistance coils and resistance boxes.—Standard B.A. unit—Subdivisions of the ohm—Combination of coils—Post-Office bridge box—

TABLE OF CONTENTS.

PAGE

Cable-testing bridge box—Dial resistance box—Thomson and Varley's slide resistance box—Precautions in using resistance coils Wheatstone's bridge — Slider bridge — Thickness of wire for resistance coils 54

Standards of electromotive force.—Post-office standard, Daniell—Warren de la Rue's chloride of silver battery—Latimer Clark's standard cell — Zinc-cadmium couple—Simple cell — Reynier's standard cell 61

Standards of capacity.—Condensers—Construction of condensers . 64

Accessory instruments.—Circuit breakers—Commutators—Reversing commutators — Reversing keys — Short-circuiting keys — Discharging keys—Double contact bridge key 66

General methods of measurement.—Direct methods—Opposition—Substitution—Comparison—Indirect methods . . . 68

MEASUREMENT OF CURRENTS.

I. GALVANOMETERS 69

Sine and tangent galvanometers—Gaugain's galvanometer—Post-Office tangent galvanometer—Schwendler's galvanometer—Siemens universal galvanometer—Thomson's reflecting galvanometer—Thomson's astatic galvanometer 70

Lamp, scale, and mirror—Hole, slot, and plane 75

Ship's galvanometer—Dead-beat galvanometers of Thomson, Marcel Deprez, Ayrton and Perry, Marcel Deprez and d'Arsonval—Torsion galvanometer of Siemens and Halske . . . 76

Ammeters and voltmeters—Ayrton and Perry's spring ammeter and voltmeter—Lieut Cardew's ammeter—Ayrton and Perry's spring and solenoid ammeter and voltmeter—Crompton and Kapp's unchangeable constant ammeter and voltmeter . . 79

Thomson's new current and potential galvanometer—Balistic galvanometer 80

Shunts and circuit resistance coils—Multiplying power—Compensation resistance 87

Constant of a galvanometer—Maximum sensibility—Theorem of sensibility—Formula of merit—Circuit resistance coils—Calibration of a galvanometer—Thickness and resistance of galvanometer wires—Shape of coils 89

Measurement of a current in C.G.S. units by the tangent galvanometer—Oscillation method 93

Indirect measurement of current strength.—By Ohm's law—By the voltmeter 94

viii   THE ELECTRICIAN'S POCKET-BOOK.

	PAGE
II. Electro-Dynamometers of Weber, Joule, and Siemens and Halske	94
III. Voltmeters	95

MEASUREMENT OF RESISTANCES.

Resistance of conductors.—By substitution—By addition to a known circuit—Wheatstone's bridge—Resistance of a conductor connected to earth—Resistance of overhead lines—Ayrton and Perry's ohm-meter—J. Carpentier's proportional galvanometer—Specific conductivity—Measurement of very high resistances—Measurement of very low resistances: Thomson's bridge 96

Resistance of galvanometers.—By half deflection—By equal deflection —Thomson's method 100

Internal resistance of batteries.—Thomson's half deflection method— By the differential galvanometer—Method when an even number of identical cells is at hand—Mance's method—Siemens' method— Munro's method. 102

Insulation of overhead lines.—Ordinary method—Determination of insulation per mile 105

MEASUREMENT OF POTENTIAL AND ELECTROMOTIVE FORCE.

Electrometers.—Electroscopes, repulsion electrometers—Thomson's absolute electrometer—Thomson's quadrant electrometer—Idiostatic and heterostatic methods—Law of deflection of the quadrant electrometer—Mascart's symmetrical electrometer—Lippmann and Debrun's capillary electrometers—Ayrton and Perry's spring electrometer 106

Indirect measurement of differences of potential.—Voltmeters— Opposition method—Partial opposition method . . . 109

Electromotive force of batteries.—Equal resistance methods—Equal deflection methods of Wiedemann, Wheatstone, Lacoine and Poggendorff—Clark's potentiometer—Law's potentiometer—By opposition 110

MEASUREMENT OF ELECTRICAL QUANTITY.

Faraday's law—Gas voltmeters—Electrolytic cells—Edison's electric meters—Coulomb meters of Edison, Ayrton and Perry— Integrating coulomb meter of Vernon Boys 116

MEASUREMENT OF CAPACITY.

Electrostatic capacity of condensers 117

MEASUREMENT OF ENERGY.

Dynamometers.—Absorption and transmission — Simple dynamometer—Ayrton and Perry's absorption dynamometer—Revolution counters and speed indicators 118

Measurement of electrical energy.—Energy expended by an electrical apparatus—Heat disengaged in a conductor through which a current is passing—Energy mètres of Ayrton and Perry, Marcel Deprez, and Vernon Boys 120

CABLE TESTING.

Standard temperature—Tank—To insulate the end of a **cable**—Instruments 122

Resistance of the conductor.—Bridge method—False **zero** method—Reproduced deflection method—Resistance of earth plates . . 123

Electrostatic capacity—Ratio of discharge—Loss of charge—Loss of half charge—Total electrostatic capacity per mile or per knot—Potential of two cables joined together—Capacity of two cables joined together 124

Insulation.—Deflection method—Differential galvanometer method —By loss of charge—Insulation of joints—Calculation of insulation—Speed of transmission—Duration of transmission—Weight of conductor and **dielectric** 126

FOURTH PART.

Practical Information.—Applications.— Experimental Results.

Algebraic formulæ.—Permutations and combinations—Newton's binomial theorem 130

Table of n; $\frac{1}{n}$; n^2; $\sqrt{n}$; n^3; $\sqrt[3]{n}$; πn; $\frac{\pi n^2}{4}$; of numbers from 1 to 100 and factors of π—Progressions and logarithms . . . 131

Table of decimal and Naperian logs from 1 to 100 . . . 135

Geometrical formulæ.—Lengths, areas, volumes—Apothemes, radii, and areas of regular inscribed polygons, in terms of the side . 137

Table of sines and tangents 139

Trigonometrical formulæ.—Solution of triangles, etc. . . . 140

Coins of different countries 144

Physical formulæ.—Fall of bodies—Moment of inertia — Formula of the bifilar suspension—Velocity of sound, light, wind, engine

	PAGE
belts, armatures, and field magnets of dynamo machines—Work produced by men and horses	144
Specific gravity of solids and liquids	148
Baumé and Cartier and Gay-Lussac's scales for liquids lighter than water	149
Baumé and Beck's scales for liquids heavier than water	151
Specific gravities of solutions of sulphuric acid in water	152
Densities of solutions of nitric acid	153
Densities of solutions of zinc sulphate and common salt	153-4
Specific gravities of gases and vapours	154
Densities of solutions of copper sulphate	155
Barometer.—Exact formula for the reduction of the height of the barometer to 0° C.—Mean height of barometer at different heights above sea level	155
Thermometer.—Fahrenheit and Centigrade thermometer scales—Determination of high temperatures	156
Linear coefficients of expansion of some solids—Cubic coefficient of expansion of mercury	157
Melting and boiling points of common bodies—Boiling points of liquids	158
Heat disengaged by the combination of one gramme of certain substances with oxygen and chlorine	159
Heat disengaged or absorbed by chemical actions—Combustion of common gas and electric light	160
Heat of liquefaction and vaporisation—Specific heat	160

RESISTANCE.

List of common materials in order of decreasing conductivity	161
Resistance of common metals and alloys at 0° C.	162
Conductivity relative to pure copper of copper alloyed with other substances, and of different samples of copper	163
Influence of temperature on the resistance of metals	164
Resistance of carbon, selenium, phosphorus, and tellurium	165
Resistance and conductivity of liquids	166
Resistance of sulphuric and nitric acids—Copper sulphate and zinc sulphate—Mixtures of copper and zinc sulphates	167
Resistance of water, ice, and glass	168
Resistance of insulators, guttapercha and indiarubber	169

CONDUCTORS.

Nature of conductors—Bare conductors—Covered wire	171
Birmingham gauges, " jauge carcasse "	172-3

	PAGE
Copper, resistance of pure copper wire	174
Weight of silk covering of wires	175
Iron—Galvanised wire—Phosphor bronze—Silicium bronze	175
Commercial types of conductors	177
Mechanical tests for insulators	178
Specific inductive capacity.—Capacities of condensers of common shapes in electrostatic units	179

MAGNETS.

Power of magnets—Supersaturation—Influence of temperature—Temper—Compressed steel—Experimental determination of the moment of inertia of a magnetised bar—To bring an oscillating magnet to rest 179

Methods of magnetising.—Single touch—Divided touch—Double touch—Elias' process—Magnetisation of a needle—Armatures for magnetised bars 181

Terrestrial magnetism.—Elements of terrestrial magnetism at Paris on Jan. 1st, 1879—At le Parc Saint-Maur on Jan. 30th, 1883 . 183

ELECTRO-MAGNETS.

Laws of electro-magnets—Maximum attraction—Action of a bar of iron in a solenoid—Formulæ for electro-magnets far from the saturation point — Electro-magnets of telegraph instruments—Construction of coils 183

Production and Applications of Electricity.

Classification.—Chemical, thermal, mechanical, and various actions 188

BATTERIES.

One-fluid cells without depolariser.—Volta's battery and its varieties—Batteries with carbon positive plates—Cells of Smee, Walker, Maiche—Cells with iron as the positive plate . . . 190

One-fluid cells with solid depolariser.—Warren de la Rue, Skrivanow, Gaiffe, Marié-Davy, and Leclanché—Oxide of copper cell of MM. Lalande and Chaperon 191

One-fluid cells with liquid depolariser.—Poggendorff, Delaurier, Chutaux, Dronier's salt, Trouvé, Tissandier 193

Two-fluid cells.—Becquerel, Daniell, Meidinger, Callaud, E. Reynier, Grove, Bunsen, Archereau, d'Arsonval's depolarising liquid—Carbons of Bunsen cells—d'Arsonval's zinc carbon cell . . 194

	PAGE
E. Reynier's jacketed zinc cells—Modifications of Grove's and Bunsen's cells—Cells of Marié-Davy, Duchemin, Delaurier—Bichromate of potash—Fuller, Cloris Baudet, d'Arsonval, Niaudet's chloride of lime battery—Circulation, agitation, and aëration	197
Thermo-chemical batteries.—Becquerel, Jablochkoff, Dr. Brard	199
E. m. f. of one-fluid batteries without depolariser, of Grove's cell, of amalgams of potassium and of zinc—Metallodion cell	199
E. m. fs. of some two-fluid cells—Theoretical e. m. fs.	202
Theoretical conditions of a perfect battery—Constants and work of some known cells—Defects of batteries—Choice of batteries according to the work they have to do—Care and maintenance of batteries—Battery testing	202

ACCUMULATORS.

Of Gaston Planté and Faure—Faure-Sellon-Volkmar accumulator—Copper and zinc accumulators of E. Reynier—Power of storage, and power of giving out of accumulators	208

CALCULATION OF ELECTRO-CHEMICAL DEPOSITS.

Chemical and electro-chemical equivalents—Calculation of the e. m. f. of polarisation in an electrolyte—Electrolysis of water—Calculation of the e. m. f. of cells—Electrolysis without polarisation	211

ELECTRO-METALLURGY.

Electrotyping.—Copper—Moulds—General management of baths and currents—Density of current—Copper clichés or electrotypes.	
Electroplating.—Coppering—Brassing—Gilding—Silvering—Silvering table plate—Nickeling	215

THERMO-ELECTRICITY.

Thermo-electric power—Inversion—Neutral point—Formula and table for the calculation of thermo-electric power—Bismuth-copper battery—Noé's battery—Clamond's battery	222

HEATING ACTION OF CURRENTS.

Loss of energy in a conductor—Heat disengaged—Limit of diameter of wires—Heating of a conductor—Heating of equal and similar coils—Electric light.—(*See* also page 260)	226

MECHANICAL GENERATORS OF ELECTRICITY.

Definitions—Work expended—Electrical energy produced—Available electrical energy—Heating of the machine—Relation between the external and internal resistances	228

	PAGE
Classification of machines—Methods of excitation—Qualities—Field magnets—Armatures—Conditions to be aimed at in a powerful machine—Influence of speed on work absorbed—Characteristic—Critical speed—Lead of brushes—Use and influence of the iron ring—Maintenance of the brushes and commutator—Thickness of wire—Working conditions	230
Continuous current machines.—A Gramme, Heinrichs, Gülcher, Schukert, Siemens, Edison, Edison-Hopkinson, Bürgin, Brush, Elphinstone-Vincent	236
Alternating current machines.—Siemens, Ferranti-Thomson, de Méritens	243
Measurement of current strength and e. m. f. of alternating current machines—Methods of MM. Joubert and Potier	245

ELECTROMOTORS.

Alternating, pole-reversing, and continuous current—Electrical work—Heating—Mechanical work—Motor driven by a battery—Deprez's and Trouvé's motors—Gramme machine with permanent magnets—Gramme and Siemens' dynamos—Ayrton and Perry's motors 247

TRANSMISSION OF POWER.

Principle—Theoretical case—Practical case—Electrical, mechanical, and commercial efficiency	251
Theoretical limit of the work transmitted by a line of given resistance	253
Gramme machines—Marcel Deprez's experiments between Miesbach and Munich, and at the Gare du Nord at Paris, and at Grenoble	255
Useful formulæ	257

ELECTRIC LIGHT.

Voltaic arc.—Classification—Monophotal and polyphotal lamps—Hand apparatus—Lamps regulated by current strength, by shunts, by differential action—Various lamps—Alternating and continuous currents	260
Resistance of the arc—Energy absorbed	262
Carbons, bare and plated	262
Gramme machines and lamps used in the French navy	264
Abdank-Abakanowicz and Gulcher lamps	265
Tests at the Paris Electrical Exhibition in 1881.—Machines and lamps of Gramme, Jurgensen, Maxim, Siemens, Bürgin, Siemens, Weston, and Brush	265-7
Electric candles.—Jablochkoff, Jamin, and Debrun	265

	PAGE
Sun lamp	268
Incandescence.—Semi-incandescent lamps of Reynier and Werdermann	269
Pure incandescent lamps of Edison, Maxim, Swan, Lane-Fox, and Siemens and Halske—High resistance lamps—Low resistance lamps—Nothomb's lamp—Bernstein lamp—Small lamps for electric jewels	270

TELEGRAPHY.

Overhead lines.—Conductors, joints, insulators—Insulation—Loss—Office wires—Earth and earth wires	274
Underground lines.—Cables—Berthoud and Borel's cables—Brook's cable	279
Submarine lines.—Table of details of the principal recently constructed cables	280
Instruments.—Classification—Visual, acoustic, registering, printing, autographic, and speaking—High speed instruments	281
Electro-magnets—Means of avoiding the extra current on breaking circuit—Strength of telegraphic currents in France and India	283
Range of electro-magnetic receivers—Sensibility—Siemens relays—Local sounders—Portable sounder	285
Dial telegraph—Morse instrument—Signals of the Morse instrument	287
Hughes telegraph—Wheatstone automatic instrument—Speed of transmission of telegraphic instruments	288

TELEPHONY.

Magnetic and battery transmitters—Receivers—Line—Induction—Losses on the line—Distance of transmission—Work of batteries in use with microphones—System of simultaneous telegraphic and telephonic transmission of Van Rysselberghe	289

FIFTH PART.
Recipes and Processes.

Alloys and amalgams.—Fusible alloys—Instrument makers' alloys—Aluminium bronze—Silvering for curved mirrors—Tombac—Romilly's brass—Nickel coins—Alloys for soldering	293
Slight metallic deposits.—To give copper the appearance of platinum—Platinised silver—Platinised carbon—Platinised iron—Amalgamation of iron and zinc—Silver black—Gilt plumbago	294
Various stores.—Cyanide of potassium—Chloride of gold—Porous pots—Morse paper—Copper sulphate—Chloride of ammonium	

TABLE OF CONTENTS.

PAGE

(sal-ammoniac)—Dextrine—Black oxide of manganese—Commercial copper and zinc sulphates—Purification of commercial sulphuric acid—Gilder's verdigris—Purification of graphite . . 296

Magnetic figures 298

Joints and soldering of wire 299

Varnish and insulators.—Red varnish—Agglomeration of wires—Coating of external wires of large electro-magnets—Cement for induction coils—Varnish for silk—Varnish for insulating paper—Insulating mixture for coils of electrical instruments—Clark's and Chatterton's compounds 300

Cements.—For insulators, Muirhead's, black, Siemens'—Marine glue—Cement to resist heat and acids—Waterproofing of wooden battery cells—Gaston Planté's cement—Cement for bone and ivory 302

Various. — Ebonite — Waterproofed vats for electro-plating—Turner's cement—Composition for rubbing the cushions of frictional electric machines—Cleaning copper and its alloys—Cleaning articles for nickel plating 303

Deposition of copper on glass—Temporary drills and tools—Black bronze—Green or antique bronze—Medal bronze—Bronzing iron—Preparation of carbon for electric light—Solution for paper of chemical telegraphs—Translation of Morse signals into letters—Fixing wires for electric house bells—Spray producer—Static induction machines—Ink for writing on glass—Ink for engraving on glass—Coppering by simple immersion 306

BIBLIOGRAPHY 312

INDEX OF TABLES 315

THE
ELECTRICIAN'S POCKET-BOOK.

First Part.

DEFINITIONS, PRINCIPLES, GENERAL RULES.

The physical phenomena, the study of which is included under the title of *electricity and magnetism*, can be subdivided into several groups, which, for want of a better classification, may be investigated in the following order:

(1) *Magnetism.*—The action of magnets on magnetic bodies, and of one magnet upon another.

(2) *Static electricity.*—The action of electrical charges.

(3) *Dynamic electricity*, or *electricity in motion.*—Laws of currents, chemical action, and heating effects.

(4) *Electro-dynamics.*—Action of currents on each other.

(5) *Electro-magnetism.*—Magnetic actions produced by currents.

(6) *Induction.*—Currents produced in closed circuits by electrical or magnetic actions outside those circuits.

We will adopt this order in the explanation of electrical laws, methods of measurement, and practical results. This classification does not, perhaps, present all necessary qualities from a scientific point of view, but it has the advantage of establishing convenient subdivisions, which facilitate research, and, to a certain extent, prevents confusion between the different subjects.

MAGNETISM.

The name *magnet* is given to all bodies capable of attracting iron. The properties of magnets, taken as a whole, and their investigation, constitute the science of *magnetism*. Magnets may be divided into three classes:

(1) *Natural magnets.*—Magnetic oxide of iron, or magnetite Fe_3O_4, or loadstone.

(2) *Artificial magnets.*—Tempered or compressed steel.

(3) *Electro-magnets.*—More or less pure iron, **magnetised by the action of a** current.

Artificial magnets are made, according to the purpose for which they are to be used, in the form of bars, needles, horse-shoes, and U's. A magnet has always at least two poles. The *axial line*, or *magnetic axis*, is the line joining the poles of the magnets; the equatorial line is that which is perpendicular to it. In a magnetised needle, the pole which turns towards the *north* is called the north pole, austral pole, marked pole, or Airy's red pole. It is indicated by the letters N or A. The pole which turns towards the *south* is called the south pole, boreal pole, non-marked or Airy's blue pole. It is indicated by the letters S or B.

Magnetic or *paramagnetic* bodies are those which, without magnetism of their own, are attracted by magnets.

Diamagnetic bodies are, on the contrary, repelled by magnets.

Laws of magnetic action.—Two poles of the same name repel each other; two poles of different names attract each other. The force exerted between two magnetic poles m and m' is proportional to the product of their intensities, and inversely proportionate to the square of the distance (d) between them;

$$f = \frac{mm'}{d^2}.$$

The *unit pole*, or *unit of magnetic quantity*, is that which, at the unit distance from a similar pole, exercises an action equal to one unit of force. The portion of space which is under the influence of a magnet is called the *magnetic field*. The intensity of the magnetic field at a given point is equal to the force which the unit pole would exert at that point. The direction of the force is that in which a pole is urged by the magnetic field, or is that which a short magnetised needle, balanced and freely suspended, would take up when placed in the field.

The space which surrounds a magnet, which is called the magnetic field, is found to be in a particular condition characterised by the presence of *lines of force*. This magnetic field is defined when we know the number of lines of force, their form, and their direction at each point of the field. These lines of force, in the simplest case (that of a magnetised bar) spread out in several directions, returning to the opposite end to that from which they started, and return through the interior of the mass of the bar. By defining a given line of force as the trajectory described by a north pole or marked pole moving freely under the influence of the magnet, the *direction* of the line of force will be: From the north pole to the south pole in the magnetic field; from the south pole to the north pole within the magnet.

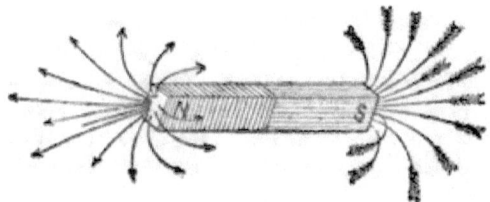

Fig. 1.—Magnetic Field.

The above sketch shows, roughly, the direction and form of the lines of force of a field produced by a magnetised bar. These lines of force have a real existence, as is shown by magnetic figures, and possess the following properties:

Properties of the lines of force.—(1) Lines of force tend to become shorter. (2) Lines of force which are parallel and in the same direction repel each other (*Faraday*). (3) A line of force passing through a magnetic body may be considered as *magnetically* shorter than a line of force of the same length passing through air. The investigation of magnetic figures confirms these theories of Faraday's in every case, and explains the mutual attractions and repulsions of magnets.

A uniform magnetic field is that of which the intensity is the same at all points, and in which the lines of force are straight, parallel, and equidistant. The magnetic action between two magnets, of which the lengths may be neglected as compared with the distance between them, is inversely proportional to the cube of the distance between them (*Gauss*).

The magnetic action between a suspended magnet and a mass acting upon it are proportional to the square of the number of oscillations which the magnet would make in a given time under their action alone, and inversely proportional to the square of the time which the magnet takes to make one complete oscillation (*Coulomb*).

Absolute moment or magnetic moment of a magnet.—Let m be the intensity of one of the poles of a magnet, and l the distance between the poles, its moment is the product ml.

Intensity of magnetisation.—The intensity of magnetisation is the ratio of the magnetic moment of a magnet to its volume.

The properties of a magnetic field may be expressed numerically by showing the intensity of the field and the direction of the magnetic force at every point. By tracing the direction of the force at each point of the field *lines of force* are obtained, and by making the *number* of these lines of force proportional to the intensity of the field at each point, a *graphic representation* of the field is obtained which is very useful in the investigation of magnetic actions and induction effects. This mode of representing a magnetic field is due to *Faraday*. When a magnetised bar, whose moment is ml, is placed in a uniform magnetic field of intensity H perpendicularly to the lines of force, a couple G is produced proportional to the intensity of the field, to that of the poles m, and to the distance between them l.

$$G = ml\mathrm{H}.$$

This couple tends to turn the needle round and cause its magnetic axis to take up a position parallel to the lines of force of the field.

Magnetic induction.—A magnetic body placed in a magnetic field is magnetised in the direction of the lines of force of the field. Its magnetism is called *induced magnetism*, and the action itself is called *magnetic induction*. The magnetism retained by a magnetic body after it has been withdrawn from the field is *residual magnetism*; the unknown cause of the residual magnetism is called *coercive force*.

Coefficient of induced magnetism, or magnetising function.—Let H be the intensity of a magnetic field, γ the intensity of magnetisation; the magnetising function k is given by the equation

$$k = \frac{\gamma}{\mathrm{H}}.$$

It is proportional for very small values of H; beyond such values k is a function of H, which diminishes when H increases, and tends towards a final value, which is called the limit of magnetisation.

TERRESTRIAL MAGNETISM.

When considering its magnetic action, the earth may be looked upon as a vast magnet, whose *marked* pole, or "north pole," is at the *south*.

Magnetic meridian.—A vertical plane passing through the magnetic axis of a magnetised needle, suspended by its centre of gravity.

Declination.—The **angle** which the magnetic meridian makes with the terrestrial meridian.*

---

* Sailors sometimes call this angle the *variation* of the compass; but this term is incorrect.

Inclination.—The angle made by a magnetised needle with the horizon in the magnetic meridian.

Intensity.—The value of the terrestrial magnetic force which is resolved into horizontal intensity and vertical intensity.

Isoclinic lines.—The locus of the points of equal inclination.

Magnetic equator, locus of points of no inclination.

Magnetic poles, points where the inclination is 90°.

Isogonic lines, locus of the points of equal declination.

Agonic lines, locus of the points of no declination.

Isodynamic lines, locus of points of equal intensity.

Variations.—Hourly, diurnal, annual, secular, etc., changes which occur in the value of the elements of terrestrial magnetism.

Magnetometers and magnetographs.—Apparatus by which the values and variations of terrestrial magnetism are measured and registered.

To neutralise the directive action of the earth on a magnetic needle.—(1) A magnetic bar is placed above the needle in the plane of the magnetic meridian, so as to act in a contrary direction to the earth. By varying its distance from the needle the action of the earth may be entirely, or in part, neutralised; the oscillations of the needle become slower as the directing force diminishes.

(2) By using *astatic needles;* two needles nearly equally magnetised and mounted on the same pivot, with their contrary poles superimposed. (*See* their use in Third Part.)

STATIC ELECTRICITY.*

Statical electricity is manifested on electrified bodies under the form of a charge. The quantity of electrification of the body gives the measure of its charge, and the nature of this charge with respect to the surrounding space determines its sign. The production of a charge of a given sign on a body always determines the production of an equal charge of opposite sign on another body.

By convention the charge taken by glass rubbed with silk is called vitreous electrification, positive (+), positive fluid, or positive electricity. The charge taken by resin, gum, indiarubber, or yellow amber rubbed with flannel is called resinous negative (—), negative fluid, or negative electricity. Bodies which show no sign of electrification are said to be in a neutral state.

---

* For a long time the name of **frictional** electricity was given to a group of phenomena produced by electrical charges. This is an improper expression, because friction is only one means (it is true, that which is most employed) for producing electrical charges.

Laws of electrical attraction and repulsion.—

Two bodies whose charges are of the same sign repel each other; two bodies whose charges are of a contrary sign attract each other. The attraction or repulsion of two charged bodies is proportional to the product of the charges, and inversely proportional to the square of the distance (*Coulomb*).

Calling the charges qq', and the distance between them d, the force f is given by the equation

$$f = -\frac{qq'}{d^2}.$$

The sign $+$ indicates an attraction; the sign $-$ repulsion.

Distribution of electrostatic charges.—

The charge of a conductor is on its surface. It is distributed uniformly over a sphere, and accumulates on points, edges, etc.

The distribution of a charge is defined by the electrical density at each point; that is to say, the quantity of electricity per unit of surface at each point.

The *potential* of a charged body is the measure of its electrification.

The electrostatic *capacity* of a body is measured by the quantity of electricity or charge which must be communicated to it to raise its potential by one unit.

The following relation exists between the potential V of a body, its charge Q, and its capacity C.

$$C = \frac{Q}{V}.$$

Electrostatic induction.—

The action exerted by a charged body on another body in the neutral state placed at a distance.

In every body in the neutral condition induction precedes attraction. Induction depends on the nature of the medium which separates the two bodies, which is called the dielectric. This influence is a measure of the *inductive capacity* of the medium.

Specific inductive capacity, or dielectric capacity.—

The ratio between the capacity of two condensers of the same dimensions, of which one is an air condenser and the other has for its dielectric the substance of which the specific inductive capacity is sought. The specific inductive capacity of dry air at 0° C. and at a pressure of 76 centimètres of mercury (*see* figures in Fourth Part) is adopted as the unit.

CONDENSERS.

Two conductors of any form separated by an insulator or dielectric, and having charges of opposite signs, form a condenser.

The Leyden jar is a condenser, so is a submarine cable. The common condensers which are used in induction coils, and as standards of capacity, are in general composed of sheets of tinfoil separated by insulating sheets (of paper, mica, etc.). These tinfoil sheets act like the internal and external coatings of a Leyden jar.

Capacity of condensers.—This capacity is measured by the quantity of electricity which the condenser contains when charged by the unit of potential. In the *electrostatic system* the units have been chosen so that the capacity of a spherical insulated conductor is numerically equal to its radius. The C.G.S. unit of electrostatic capacity is the capacity of an insulated spherical conductor of one centimètre radius. In the *electro-magnetic* system (the only one which is employed in practice) the unit is the *farad*, a condenser which, when charged to the potential of one volt, contains one coulomb of electricity. In practice the microfarad is most commonly used. (*See* Second Part.)

Charge of a condenser.—The charge Q taken by a condenser is equal to the product of its capacity C by the e. m. f. E by which it is charged.

$$Q = CE.$$

Example.—A condenser of 0·5 of a microfarad charged to a potential of 150 volts would contain 0·5 × 150 = 75 microcoulombs of electricity.

Charge taken by two condensers.—Two condensers of capacity C and C′, connected one to the (+) pole, the other to the (−) pole of an insulated battery, their other armatures being to earth, take equal charges of a contrary sign (+ q and − q). The relations between the charges, the capacities and potentials v and v', are the following, E being the e. m. f. of the battery:

$$q = cv \qquad q' = c'v' \qquad v - v' = E.$$

$$v = \frac{Ec'}{c+c'} \qquad v' = \frac{Ec}{c+c'} \qquad q = \frac{E}{\frac{1}{c}+\frac{1}{c'}}.$$

Condensers joined up for surface.—Let $a\ b\ c \ldots$ be the individual capacity of each condenser, the total capacity $C = a + b + c + \ldots$

If the condensers are charged separately with quantities,

$$q = av \quad q' = bv' \quad q'' = cv'';$$

the total charge Q, when they are joined up for surface, will be

$$Q = q + q' + q'' \ldots$$

Their common potential V will be

$$V = \frac{Q}{C} = \frac{av + bv' + cv'' \ldots}{a + b + c \ldots}$$

If one of the condensers has a charge $q = a\,v$, after they are joined up the common potential will be

$$\frac{q}{c} = \frac{av}{a + b + c \ldots}$$

Condensers joined up in cascade.—Internal armature of the first joined to the source E, external armature to internal armature of the second, external armature of the second to internal armature of the third, etc.; the external armature of the last to earth.

The *capacity* of the system is given by the relation

$$\frac{1}{C} = \frac{1}{a} + \frac{1}{b} + \frac{1}{c} \ldots$$

if the *n* condensers have the same capacity *a*,

$$V = \frac{nQ}{a}, \quad Q = \frac{aV}{n}, \quad C = \frac{a}{n}.$$

If the system be discharged, the quantity of electricity which traverses the external circuit is

$$\frac{aV}{n}.$$

But if the condensers be separated, each of them separately has a charge of the same value.

Energy of condensers.—The energy due to the discharge of a condenser has for its value

$$W = \frac{1}{2}\,QV = \frac{1}{2}\frac{Q^2}{C} = \frac{1}{2}\,V^2 C;$$

Q being the charge, V the potential, C the capacity. When V is expressed in volts, C in farads, and Q in coulombs, the energy of the condenser in ergs, in kilogrammètres, or in calories is

$$W = \frac{1}{2} V^2 C \text{ ergs};$$

$$W = \frac{1}{2} V^2 C \times \frac{1}{98 \cdot 1 \times 10^6} \text{ kgm.} = \frac{1}{2} V^2 C \frac{7 \cdot 2331}{98 \cdot 1 \times 10^6} \text{ foot-pounds};$$

$$W = \frac{1}{2} V^2 C \times \frac{1}{42 \times 10^6} \text{ calories (g.-d.)}.$$

These relations are made use of in methods of measurement founded upon the use of condensers, testing of submarine cables, and the magnificent experiments made by M. Gaston Planté with his rheostatic machine.

Contact electricity.—The contact of two bodies of different kinds produces a difference of potential between them. This difference of potential varies with the bodies in contact.

Volta's law.—The difference of potential between two metals is equal to the algebraic sum of the differences of potentials due to the contact of the intermediate metals.

DYNAMIC ELECTRICITY, OR ELECTRICITY IN MOTION.

LAWS OF CURRENTS.

When two points at different potentials are joined by a conductor, a flow of electricity passes along the conductor, which joins these two points; this flow is called a *current*. If the points joined by the conductor are only connected to bodies charged with a certain quantity of electricity, the flow will only last for an instant, and will constitute a *discharge*; if by any means the difference of potential is kept constant a true current is obtained; the cause which produces the current is called *electromotive force*, and any apparatus in which it is developed constitutes a *generator of electricity*.* The greater or less opposition which a conductor opposes to the current is the *resistance* of the conductor; the strength of the current is equal to the quantity of electricity which traverses the conductor in one unit of time. The current is the same at

---

* In some text-books it has been called an "electromotor," but this term at the present day means an instrument for converting the energy of an electric current into mechanical work

all points of the circuit. The current strength, the electromotive force, and the resistance are connected by Ohm's law.

Ohm's law.—The strength of a current is proportional to the electromotive force, and inversely proportional to the resistance of the circuit. Calling the strength of the current C, the electromotive force E, and the resistance R, Ohm's law is thus expressed:

$$C = \frac{E}{R}$$

Kirchoff's laws.—(1) At every point of junction, that is to say, at every point where two or more conductors join, the sum of the strengths of the currents is zero, considering the currents which flow towards the point as positive (+), and those which flow away from it as negative (−).

(2) In every closed system of conductors the sum of the products of the current strengths by the resistances is equal to the sum of the electromotive forces, considering those positive which produce an increase of potential, and as negative those which produce diminution of potential.

Bosscha's corollaries.—(1) When in a system of closed circuits the strength of the current is zero in one of the branches, the current strengths in the other branches are independent of the resistance of the conductor in which there is no current.

This resistance may vary from zero to infinity without affecting the rest of the system.

(2) When two branches, A and B, of a system or network of conductors are such that an electromotive force placed in branch A sends no current into branch B, the resistance of A may be varied from zero to infinity without disturbing the condition of branch B.

Specific resistance of a substance.—The specific resistance of a substance is the value in absolute units of the resistance of a cube of this substance having for side the unit of length; this resistance being measured between two opposite faces.

Conductivity.—This is the reciprocal of resistance. This is very little used now in practice except as a means of estimating the relative value of electrical conductors. For this purpose the conductivity of pure copper at 0° C. is taken as the standard. It is represented by 100 or by 1.

Resistance of a conductor.—The resistance R of a conductor is proportional to its length l, and inversely proportional to its

RESISTANCE—BATTERIES.

sectional area s, and proportional to the specific resistance a of the substance of which it is made.

$$R = \frac{al}{s}.$$

When the conductor is cylindrical, its resistance is therefore inversely proportional to the square of its diameter d.

$$R = \frac{al}{d^2} \times \frac{4}{\pi}.$$

Resistance of derived, branch, parallel, or shunt circuits.—When there are two derived circuits, a and b, their united or reduced resistance R is equal to their product divided by their sum.

$$R = \frac{ab}{a + b}.$$

When there is *any number* of circuits, calling their resistances $a\ b\ c$, their united resistance R is equal to the reciprocal of the sum of their reciprocals:

$$R = \frac{1}{\frac{1}{a} + \frac{1}{b} + \frac{1}{c} + \ldots}$$

When the n derived circuits are all of them of the *same resistance* a, we have for their united resistance the expression

$$R = \frac{a}{n}.$$

BATTERIES.

A battery is an apparatus which produces electricity by chemical action, generally by the oxydation of zinc and sometimes of iron. A battery reduced to its simplest terms is called a *cell* or *element*; several such elements joined together are spoken of usually as a *battery*. A battery, or, rather, a cell, is composed generally of two metallic plates, immersed either in one liquid or in two different liquids. The points at which the exterior conductors are attached are called *poles* or *electrodes*. The oxydised plate (generally zinc) forms the negative pole; the other, or reduced plate, the positive pole. By convention we suppose that the development of electrical energy produced by chemical action is manifested under the form of a flow or current of electricity, of which the direction is defined by saying that in the external circuit the current produced goes from the positive pole (+) to the negative pole (—), and in the element itself, from the negative pole to the positive pole. This convention, which is convenient for the explanation of the phenomena, in no way prejudges the real nature of the current, about which up to the present time we know nothing.

In all batteries actually in use zinc is the negative pole; platinum, carbon, or copper, the positive pole.*

Elements are said to be put up for *tension* or in series, when the + pole of the first is joined to the − pole of the second, the + of the second to the − of the third, and so on.

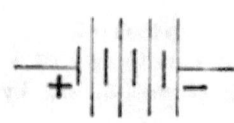

Fig. 2.—Conventional Representation of a Battery.

Elements are said to be put up for *quantity*, for *surface*, or *parallel*, when all the positive poles of the battery are joined together, and all the negative poles together. When they are put up, so that the e. m. f. of one opposes that of the other, they are said to be put up in *opposition*.

Conventional representation of a battery.—To avoid the repetition of an actual drawing of a battery, the conventional sign represented at the side (Fig. 2) is used in diagrams. The fine lines represent the zincs (−), and the thick lines the coppers (+). Their number sometimes indicates how many elements there are.

Laws of the chemical actions in batteries.—The quantity of chemical action, or the quantity of zinc dissolved in a battery, is theoretically proportional to the quantity of electricity which it produces, and the quantity of chemical action per unit of time is proportional to the strength of the current. In a battery of elements put up in series, the quantity of chemical action is the same in each element.

Constancy of batteries.—The electromotive force of a battery depends on the nature of the chemical reactions of the substances employed, their concentration and temperature. The internal resistance depends on its form and dimensions. It may be diminished by bringing the plates near together, and by increasing their dimensions. The electromotive force E of a battery, and its internal resistance r, are called its *constants*. A battery is said to be constant when its constants do not change during its action.

The polarisation of a battery is the falling off of its electrical energy;

---

* On account of the conventional direction of the current within the element, the negative plate is sometimes called the electro-positive plate; and the positive plate, the electro-negative plate. These terms are often met with in the older text-books, particularly in discussing the electrical relations of different metals and solutions to each other.

it is due to the deposit of hydrogen on the positive plate, which increases the internal resistance and diminishes the e. m. f.

The polarisation may be overcome by agitation, blowing air through the liquids, the employment of rough battery plates, or by the employment of systems in which the gaseous hydrogen is replaced by a deposit of solid metal (Daniell's battery), or by the employment of a second liquid surrounding the positive plate, capable of combining with the hydrogen (Grove's and Bunsen's cells).

Energy of a battery.—Two batteries may generally be compared with one another by placing them in analogous conditions of action. The most simple method is to measure the strength of the current C which the element can give when doing its maximum work, that is to say, with an external circuit equal in resistance to the internal resistance of the battery (supposing the battery to be constant); calling E and r the constants of the elements, we have for C_m

$$C_m = \frac{E}{2r}.$$

Sometimes the value of the maximum available work, W_u, is also given, which is calculated (in kgms.) by the formula

$$W_u = \frac{EC_m}{4g} = \frac{E^2}{4rg}.$$

This is the maximum useful rate of work of the battery in kilogram-mètres of electrical energy per second (E in volts, C_m in ampères, r in ohms, $g = 9.81$). The total maximum rate of work (W_t) is double this value:*

$$W = \frac{E^2}{2rg}.$$

When a battery is working on an external circuit equal to its internal resistance, the useful available work in the exterior circuit is equal to half the total energy furnished by the chemical action. The efficiency, therefore, is 50 per cent.

Arrangement of batteries.—We will suppose that all the elements are identical. The constants of one element are, E for e. m. f., r for internal resistance.

---

* These values may be obtained in foot-pounds per second by the following formulæ:

$$W_u = \frac{EC_m}{2 \times 1.35} = \frac{E^2}{4r \times 1.35}; \quad W_t = \frac{E^2}{2r \times 1.35}.$$

(1) n elements arranged in series behave like one single element, of which the e. m. f. $= n\,E$, and the internal resistance $= n\,r$.

(2) n elements put up for quantity or parallel behave like one single element of the same e. m. f. E, and of which the internal resistance $= \dfrac{r}{n}$.

(3) n elements arranged thus, t in series, q parallel, behave like a single element, of which the e. m. f. $= tE$, and the internal resistance $= \dfrac{t}{q} r$.

Maximum effect.—To obtain the maximum effect of n elements on an external circuit of resistance R, the internal resistance of the battery must be equal to the external resistance R; that is to say,

$$\frac{t}{q} r = R, \qquad (1)$$

with the condition that the ratio $\dfrac{t}{q}$ must be a whole number. The equation (1) combined with the equation $n = tq$ enables us to calculate the values of t and q. The strength C of the current is then

$$C = \frac{tE}{\dfrac{t}{q} r + R} = \frac{tE}{2R}.$$

These formulæ apply also to continuous current magneto-electric machines, or separately-excited continuous current dynamo machines.

Shunted battery: Pollard's theorem.—Let E and r be the constants of a battery. When this battery, which is supposed to be constant, is shunted by the resistance s, it behaves like a new battery, of which the e. m. f. is $\dfrac{Es}{r+s}$ and the internal resistance $\dfrac{rs}{r+s}$.

ELECTROLYSIS.

Laws of electrolysis.—*Electrolysis* is the decomposition of liquids, or solutions, produced by a current. Bodies which are thus decomposed are called *electrolytes*. The two poles immersed in the liquid to be electrolysed are called *electrodes*; that which is connected to the positive pole of the battery is the *anode*; the other is the *cathode*. Faraday called the bodies produced by electrolysis *ions*. Those which go to the anode are *anions*, and those which go to the cathode are *cathions*.

Faraday's laws.—An elementary body cannot be an electrolyte. Electrolysis does not take place in solids. The quantity of electro-chemical action is the same at all points of a circuit. The quantity of an ion liberated from an electrolyte in unit of time is proportional to the strength of the current. The quantity of an ion liberated at an electrode per second is equal to the strength of the current multiplied by the *electro-chemical equivalent* of the ion, and reciprocally the quantity of electricity which has passed through the electrolyte in a given time is equal to the weight of the ion which has been liberated, divided by the electro-chemical equivalent of the ion.

The electro-chemical equivalent of a body is the quantity of this substance liberated by the passage of a unit quantity of electricity. The electro-chemical equivalent is proportional to the chemical equivalent.

HEATING ACTION OF CURRENTS.

Joule's law.—The quantity of heat H disengaged in a conductor is proportional to the resistance of the conductor, to the square of the current strength C, and to the time t during which the current passes. We have then

$$H = \frac{1}{A} RC^2 t,$$

A being the mechanical equivalent of heat. When combined with Ohm's law, Joule's law also takes the two following forms:

$$H = \frac{1}{A} EC t = \frac{1}{A} \frac{E^2}{R} t,$$

E being the difference of potential between the two extremities of the resistance R.

Calling Q the quantity of electricity which traverses a conductor in the time t, by **Faraday's law** $Q = Ct$. Joule's law is also written thus:

$$H = \frac{1}{A} QE.$$

Numerical relations.—When the current strengths C are expressed in ampères,[*] the electromotive forces E in volts, the resistances in ohms, and the quantity of electricity Q in coulombs, the quantity of heat in calories (g.-d.) is expressed by

$$H = \frac{C^2 R t}{4 \cdot 16} = \frac{QE}{4 \cdot 16} \text{ calories (g.-d.)}.$$

(*See* Thermo-electricity in Fourth Part.)

---

[*] See Second Part for units of electrical measurement.

WORK PRODUCED BY CURRENTS.

Joule's law.—The work W equivalent to the passage of a current in a conductor is expressed by

$$W = C^2 R t = E C t = \frac{E^2}{R} t$$

By Faraday's law $Q = Ct$.
Joule's law may therefore be written thus:

$$W = QE.$$

Numerical relations.—When the current strengths C are expressed in ampères, the electromotive forces E in volts, the resistances R in ohms, and the quantities of electricity Q in coulombs, the work is expressed by

$$W = 10 C^2 R t = 10 E C t = 10 \frac{E^2}{R} t \text{ meg-ergs.}$$

$$W = C^2 R t = E C t = \frac{E^2}{R} t \text{ watts.}$$

$$W = \frac{C^2 R}{9 \cdot 81} t = \frac{EC}{9 \cdot 81} t = \frac{E^2}{9 \cdot 81 R} t \text{ kilogrammètres.}$$

$$W = \frac{C^2 R}{1 \cdot 356} t = \frac{EC}{1 \cdot 356} t = \frac{E^2}{1 \cdot 356 R} t \text{ foot-pounds.}$$

The work per second is obtained by making $t = 1$ in the preceding equations.

The work done is obtained in terms of the quantities of electricity by these formulæ:

$$W = 10 QE \text{ meg-ergs.}$$

$$W = QE \text{ watts.}$$

$$W = \frac{QE}{9 \cdot 81} \text{ kilogrammètres.}$$

$$W = \frac{QE}{1 \cdot 356} \text{ foot-pounds.}$$

Electrical or galvanic field.—The space surrounding a conductor conveying a current. A galvanic field is characterised by a kind of whirl-form of the lines of force. The lines of force are circles concentric with the current; their number is proportional to its strength.

In the case of a rectilinear current, when we look at the end of the conductor at which the current enters (+), the direction of the lines of force is that of the hands of a watch. The lines of force of a galvanic field possess the same

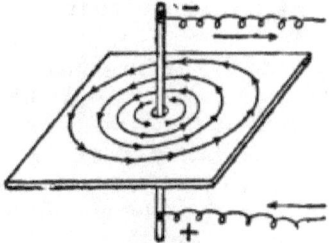

Fig. 3.—Galvanic Field.

properties as the lines of force of a magnetic field. By considering them, Ampère's laws on the mutual action of currents may be deduced.

Magnetic shell.—A closed circular current is analogous to a sort of plate magnet or magnetic shell, of which one of the surfaces is north and the other south. The following rule enables us to determine the polarities of each surface. When, following the usual convention, the current circulates in the direction of the hands of a watch, the surface at which we are looking constitutes the south pole of the magnetic shell; on the contrary, it is the north pole if the current circulates in the opposite direction to the hands of a watch. In the centre of the circular current the lines of force are perpendicular to the plane of the current.

Solenoid.—A series of circular currents forms a solenoid. Their

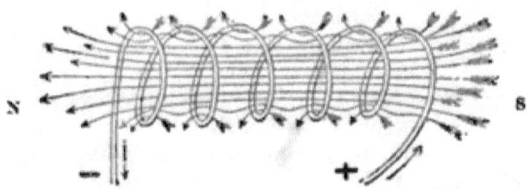

Fig. 4.—Solenoid.

reciprocal actions modify the direction of the lines of force, and cause them to assume the distribution indicated by the sketch. Thus a solenoid is fairly analogous to a *magnet*, of which the north (marked) pole is that at which, when the extremity is looked at, the current circulates in the reverse direction to the hands of a watch.

ELECTRO-DYNAMICS.

Mutual action of two currents (*Ampère's laws*).—Two parallel currents flowing in the same direction attract each other; two currents flowing in opposite directions repel each other. The force exerted between two parallel currents is equal to the products of the strength of the currents multiplied by their length, and divided by the square of their distance.

Two parts of the same current repel each other.

Two currents forming an angle with each other attract when they both approach or both leave their point of crossing; they repel each other if one of them approaches and the other flows away from the point of crossing: thus they tend to place themselves parallel to each other. A sinuous current produces the same effect as a rectilinear current terminating at the same extremities.

Mutual action of two very short currents (*Ampère's formula*).—Two very short conductors of lengths $ds\ ds'$, conveying currents of current strength cc',* attract or repel each other in the direction of the line joining their centres with a force f.

$$f = \frac{cc'\ ds\ ds'}{r^2} \left(\cos \omega - \frac{3}{2} \cos \alpha \cos \alpha'\right).$$

r being the distance between their centres, ω the angle formed by the two short conductors, α and α' the angles made by them in the one case with the line joining their centres, and in the other with its prolongation.

When f is positive an attraction is indicated, when f is negative, a repulsion.

The formula may also be written in the form

$$f = \frac{cc'\ ds\ ds'}{r^2} \left(\sin \alpha \sin \alpha' \cos \theta - \frac{1}{2} \cos \alpha \cos \alpha'\right),$$

θ being the angle between the planes passing through the two short conductors, a the line joining their centres.

English formula.—When the current strengths are expressed in electro-magnetic units, Ampère's formula is written,

$$f = \frac{CC'\ ds\ ds'}{r^2} \left(2 \cos \omega - 3 \cos \alpha \cos \alpha'\right).$$

If $ds\ ds'$ and r are expressed in centimètres, and C and C' in C.G.S. units of current strength, f is given by this formula in *dynes* (*see* page 36).

---

* The notation cc' expresses electro-dynamic units, and the notation CC' electro-magnetic units. The relation between the two is $c = \sqrt{2}\ C$.

MUTUAL ACTION OF TWO VERY SHORT CURRENTS.

Two rectilinear parallel currents.—If one is of finite length l, and the other very long in comparison, and d be the distance between them,

$$f = \frac{cc'\, l}{d},$$

and in electro-magnetic units,

$$f = \frac{2CC'\, l}{d}.$$

Two plane circuits at right angles at distance d.—One of the circuits is fixed and the other free to move. The moment of the couple tending to turn the free circuit is,

$$M = \frac{ss'\, cc'}{d^3};$$

s and s' being the areas of the circuits, and cc' the strengths of the currents passing through them in electro-dynamic units. The moment is given in electro-magnetic units by the formula,

$$M = \frac{2\, ss'\, CC'}{d^3}.$$

Action of two coils of n and n' turns of wire at a distance d.—The moment of the couple M exerted between the fixed and the movable coil is,

$$M = \frac{nn'\, ss'\, cc'}{d^3};$$

and in C.G.S. electro-magnetic units,

$$M = \frac{2nn'\, ss'\, CC'}{d^3};$$

ss' being the mean areas of the coils, and d the distance between the centres of the coils, which is supposed here to be great as compared with their dimensions.

Action of the earth on currents.—The earth exercises a directive action on currents analogous to that which would be produced by a continuous current passing round the equator, and flowing from east to west. This hypothetical current also explains the directive action of the earth on a magnetised needle.

Astatic conductors.—Conductors wound one on the other so as to destroy the directive action of the earth, used in experiments on the mutual action of currents.

Solenoids.—A series of equal circular currents in the same direction whose planes are perpendicular to the line passing through the centre of all the circles, whether this line be straight or curved.

Properties of solenoids.—A solenoid places itself north and south under the action of the earth; the end at which the current circulates in the direction of the hands of a watch points to the south, the opposite end to the north. The end which points to the north is the north pole (marked pole), the other the south pole. The poles of the same name of two solenoids repel each other, the poles of different names attract each other; the same actions are produced between a magnet and a solenoid.

COMPARISON OF MAGNETS TO SOLENOIDS.

In order to simplify the explanation of the phenomena we may compare a magnetised body to a series of juxtaposed files of circular currents, or to a bundle of solenoids. Solenoids and magnets, then, behave in the same way. By replacing a magnet by a solenoid we may explain all the actions of magnets one on the other, and of currents on magnets. We may remember the direction of these *Ampère's currents* in a magnet by remembering that when we look at the marked end (north or austral pole), the currents circulate in a contrary direction to the movement of the hands of a watch. These currents are sometimes called Ampère's molecular currents.

ELECTRO-MAGNETISM.

Fundamental principles.—When a current passes through a wire placed parallel to a magnetised needle, which is free to move, it deflects it through a certain angle which increases with the strength of the current.

Ampère's rule for the action of currents on a magnetised needle.—If we suppose an observer lying on the wire which the current passes through, in such a position that the current goes in at his feet; if he looks at the needle, he will see the north, austral, or marked pole, of the magnetic needle deflected towards his left hand. Supposing the current to have a right and left side, we say that the north pole of the magnet is always carried to the left of the current.

Multiplier.—When the wire makes several turns round the needle the action of the current is multiplied. The system constitutes a multiplier. Galvanometers (*see* Second Part) are applications of the principle of the multiplier.

The deflection produced by a current on a magnetic needle is independent of the magnetic intensity of the needle. When a circular current placed in the magnetic meridian deflects a magnetised needle, the length of which is infinitely small in proportion to the diameter of the circle, the strength of the current is proportional to the tangent of the angle of deflection (*Weber*). This is the principle of the tangent galvanometer. When the coils are turned so that the needle is again parallel to the turns of wire the strength of the current is proportional to the sine of the angle through which the coil is turned; here the law is strictly followed, whatever may be the dimensions of the needle and the form of the coil. This is the principle of the sine galvanometer.

Electro-magnet.—By introducing a bar of iron into a solenoid the lines of force developed by a current traverse the bar, and transform it into an electro-magnet, of which the power depends upon the strength of the current, the number of turns in the solenoid, etc. These magnetic properties last as long as the current is passing, and cease immediately it is interrupted.

Action of a magnet on a magnetic shell.—It is easy to foresee the reciprocal action of a magnet and a circular current by referring to the properties of the lines of force. All possible cases are comprised in an elegant rule due to Clerk Maxwell.

Clerk Maxwell's rule.—When a magnet is in the presence of a circuit, each portion of the circuit acts upon the magnet in such a direction as would cause the magnet, were it free to move, to take up the position in which the greatest possible number of its lines of force would be embraced by the circuit. From this rule it follows that if a magnet is free to move there will be movement, attraction, or repulsion, according to the relative position of the lines of force. This rule leads us rapidly to the phenomena of induction, and we are now in a position to undertake their investigation, if what has gone before has been well understood.

Action of a short conductor conveying a current on a magnet pole (*Ampère's formula*).—Let ds be the length of the conductor, m the intensity of the pole, r the distance between the pole and the conductor, C the strength of the current.

Direction.—The force exerted between the conductor and the pole is in a direction normal to the plane passing through the pole and the conductor.

If we imagine an observer lying in the conductor so that the current enters at his feet and so that he looks towards the pole, the pole will be urged from his right to his left if it be a *north* pole, and from his left to his right if it be a *south* pole.

Magnitude.—The force f exerted is given by this equation,

$$f = \frac{m\, C\, ds\, \sin a}{r^2}.$$

It is proportional to the intensity of the pole, to the current strength, and to the length of the conductor, and inversely proportional to the square of the distance between the pole and the conductor.

Galvanic field.—We have already stated that the space surrounding a conductor conveying a current is called a galvanic field. It is characterised by lines of force analogous to those of a magnetic field, but which are only in existence whilst the current is passing.

The galvanic field produced at any point is defined by the direction and magnitude of the force exerted on a unit north pole placed at the point.

In the case of a circuit of given **form** the field is determined by the resultant of all the actions of the circuit split up into infinitely short elements. When the circuit is of some simple geometrical form the field produced by it can be easily determined. We will give the formulæ for the simpler and more important forms.

Arc of a circle.—Let l be its length, and r its radius, and C the strength of the current; the intensity of the field at its centre will be,

$$f = \frac{Cl}{r^2}.$$

The force is perpendicular to the plane passing through the arc and its centre.

Circle.

$$f = \frac{2\pi C}{r}.$$

Circle of n turns of wire.—Calling the mean radius r,

$$f = \frac{2\pi n C}{r}.$$

Intensity of the field produced by a circle of radius r on the perpendicular to its plane, passing through its centre at a distance d,

$$f = \frac{2\pi C r^2}{(r^2 + d^2)^{\frac{3}{2}}}.$$

Putting $\pi r^2 = S$ (area of circle of radius r) and $r^2 + d^2 = \rho^2$, we get,

$$f = \frac{2CS}{\rho^3}.$$

SOLENOID.

Solenoid.—*Intensity of the field on the axis.*—Let n be the number of turns of wire, r the radius, and $2l$ the length of the solenoid, and a the distance of the point M on the axis from the nearest end of the solenoid; then,

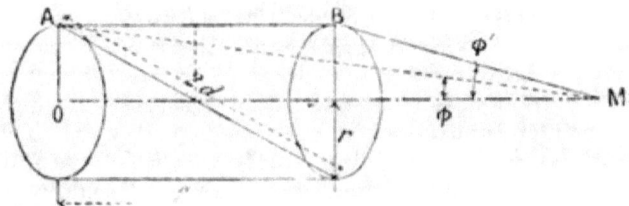

Fig. 5.—Solenoid.

$$f = \frac{\pi n C}{l}\left(\frac{a+2l}{\sqrt{r^2+(a+2l)^2}} - \frac{a}{\sqrt{r^2+a^2}}\right).$$

Calling the angle AMO ϕ, and the angle BMO ϕ',

$$f = \frac{\pi n C}{l}(\cos\phi - \cos\phi').$$

Intensity of the field at the centre.—The intensity of the field is at a maximum at this point,

$$f_c = \frac{2\pi n C}{\sqrt{r^2+l^2}}.$$

Calling the diagonal of the solenoid $2d$,

$$f = \frac{2\pi n C}{d}.$$

Intensity of the field produced by an infinitely long rectilinear current.—Let r be the distance of the point under consideration from the current,

$$f = \frac{2c}{r}.$$

Intensity of the field produced by a plane closed circuit.—First, at a point on the normal passing through the centre of gravity of the area of the closed circuit, and at distance d,

$$f = \frac{2SC}{d^3},$$

s being the area of the circuit.

Second, at a point in the plane of the circuit at distance d from its centre,
$$f = \frac{SC}{d^3}.$$

Magnetisation by currents.—A wire traversed by a current possesses temporary magnetic properties. When an insulated wire is wound a great number of times round a piece of soft iron, the passage of a current develops a powerful magnetisation, which ceases when the current is interrupted, if the iron is very soft. This is an *electro-magnet*, of which the power varies with the dimensions, the strength of the current, and the number of turns of wire, etc. (*See* Fourth Part.) When the iron is not perfectly soft, it retains some residuary magnetism when the current is interrupted.

Rule for finding the poles of an electro-magnet.—When a current passes through the wire of the electro-magnet, if we look at the end of each pole, the south pole is that one at which the current circulates in the direction of the hands of a watch, and the north pole (marked pole) that at which the current circulates in the reverse direction. This is in accordance with Ampère's hypothesis of molecular currents.

INDUCTION.

Every relative displacement of a conductor and a magnetic or galvanic field produces an electromotive force in the conductor, and if the conductor form part of a closed circuit, an induced current is set up in it in consequence.

A magnet producing a magnetic field is in this case called a *field magnet*, and a circuit producing a galvanic field is called a *primary circuit*. The conductor in which the current is induced is called an *armature* if the induction be produced by a magnet, and a *secondary circuit* if the induction be produced by a current. Induced currents may be produced either by relative *mechanical displacement* of the inducing field and the conductor (magneto- and dynamo-electric machines) or by *variation* of the galvanic field produced by the current passing through the primary circuit (induction coils). The induction of a current on its own conductor is called *self-induction*, and the current induced in a conductor by making or breaking circuit is called the extra current.

Induction produced in a rectilinear circuit displaced parallel to itself in a uniform magnetic field.—*Electromotive force of induction.*—Let e be the e. m. f.

due to induction, H the intensity of the magnetic field, l the length of the rectilinear circuit, v its velocity, α the angle made by the conductor with the direction of the lines of force, ϕ the angle between the direction of motion and the direction of the force exerted between the magnet field and the circuit; then,

$$e = Hlv \sin \alpha \cos \phi.$$

If the conductor be moved in the direction of the force,

$$\phi = 0 \quad \therefore \quad \cos \phi = 1,$$

and

$$e = Hlv \sin \alpha.$$

Energy of induction.—Let t be the time in which the displacement takes place, and W the energy of induction,

$$W = \frac{H^2 l^2 v^2 t \sin^2 \alpha \cos^2 \phi}{R},$$

R being the total resistance of the circuit.

Direction of the induced current.—If we imagine an observer to be lying in the field so that the lines of force pass in at his feet, and facing in the direction of the movement of the conductor, the current produced would pass from his left to his right.

Induction in a closed circuit.—When the whole circuit is moved, if there be no variation in the number of lines of force included in the area enclosed by the conductor forming the circuit during its displacement, there is no induced current, because the e. m. f. developed in each element of the circuit at any given moment is balanced by an equal e. m. f. of opposite sign developed at the same moment in the symmetrically corresponding element.

When the number of lines of force included by the circuit varies, the current produced can be easily deduced from Maxwell's law. When the displacement is such that the circuit tends to include a *greater* number of lines of force the induced current is in the *inverse* direction to that of a current which, if passing through the circuit, would produce motion in that direction. When the displacement is such as to cause the circuit to include a *smaller* number of lines of force, the induced current is direct, that is to say, is in the same direction as a current which would produce motion of the circuit in that direction.

Take, for example, a simple case, that of a magnet situated perpendicularly to a at a certain distance from a current passing through a conductor bent into the form of a circle.

In order that the magnet may be attracted in the position represented in the figure, *i.e.* in order that the magnet may be displaced in the direction indicated by the dotted arrow, a current must circulate in the conductor in the direction shown by the full line arrows. If the magnet be displaced by an *external mechanical force* in the direction shown by the dotted arrow, it may be observed, by means of a galvanometer G, that the current induced is in the *inverse* direction to that which would produce this displacement of the magnet. This induced current is produced because the circuit tends to include a *larger* number of lines of force. In the same way it may be observed that when the circuit tends to exclude a *smaller* number of lines of force, *i.e.* when the circuit and magnet are separated, the induced current is *direct*, *i.e.* in the same direction as the current which, passing through the circuit, would tend to produce the same movement.

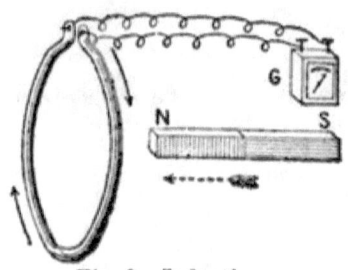

Fig. 6.—Induction.

It may thus be seen that the terms *direct* and *inverse* are relative: a *direct* current is produced by a diminution in the number of lines of force included by the circuit, and an *inverse* current by an increase in the number of lines of force. The strength of the induced current depends on the number of lines of force passed over by the circuit at each instant, *i.e.* it increases with the power of the magnet and the velocity of the displacement. If the circuit has an independent current passing through it, the induced current increases or diminishes it according as it is direct or inverse. It may be deduced directly from Maxwell's law, that a magnet approaching a circuit produces an *inverse* current, and a magnet receding from a circuit a *direct* current. A current approaching a circuit, beginning to flow, or increasing in strength, produces an *inverse* current; a current receding, ceasing to flow, or diminishing in strength, produces a *direct* current.

Law of induced currents.—The general laws of induction are summed up in the following table, and expressed in accordance with the doctrine of the conservation of energy, by Lenz's law:

Inductor.	Inverse Induced Current.	Direct Induced Current.
A magnet	Approaching.	Receding.
A current	Approaching. Beginning. Increasing in strength.	Receding. Stopping. Diminishing in strength.

Extra current.—The induced current produced by the induction of a current upon itself is called an *extra current*. The closing of the circuit produces an inverse extra current; the breaking of a circuit, a direct extra current.

Influence of extra currents on induced currents.—A current which is commencing being resisted by the inverse extra current, develops an induced *inverse* current, which lasts for some little time, and is of little strength. A current which is interrupted develops a direct induced current, of short duration, but of considerable strength. The quantity of electricity produced in each case is the same, the currents only differ by the length of time for which they last.

Induction coils, magneto-electric generators, and magneto-telephone transmitters are the most important applications of the phenomena of induction.

Lenz's law.—When a circuit is moved in the presence of a current or magnet, or a magnet is moved in the presence of a current, the induced current is such that it tends to stop the movement.

Corollaries from Lenz's law.—In a circuit under the influence of induction, or secondary circuit, the strength of the current developed is proportional to the strength of the current in the primary circuit. The electromotive force developed in a coil by the induction of a magnet is, other conditions remaining the same, proportional to the number of turns of the wire. It is independent of the diameter of the coil and the conductivity of the wire, but for the same number of turns of wire the strength of the current is inversely proportional to the resistance of the wire, and consequently inversely proportional to the diameter of the coil.

Conservation of energy in induction.—Lenz's law, by its completeness and simplicity, unites the reciprocal action of magnets and currents to the principle of conservation of energy in a striking manner, by stating that the current induced in each case tends to *resist* the corresponding mechanical motion. Mechanical generators of electricity and electromotors are striking confirmations of this truth. In both a quantity of mechanical work expended or produced always corresponds to an equivalent quantity of electrical energy produced or absorbed. Electrical energy is therefore like heat, *a mode of motion* of which the real nature is unknown, but its manifold transformations prove that it does not lie outside of the great principle of the *unity of the physical forces*.

Second Part.

UNITS OF MEASUREMENT.

Most of the quantities which have to be dealt with in physical science may be expressed in terms of *three* units, which are called *fundamental units*. All the others are deduced from them by definition, and are called *derived units*. In theory the three fundamental units may be any arbitrary *length*, *mass*, and *time* ; in practice, in the system established by the British Association, and adopted by the International Congress of Electricians in 1881, the three following units have been chosen:

> *Centimètre*.—Unit of length.
> *Gramme*.—Unit of mass.
> *Second*.—Unit of time.

The system of units based on these three fundamental units is called the *centimètre - gramme - second* system, or, by abbreviation, the C. G. S. system.

Fundamental units.—The three fundamental units are expressed by symbols: L for the unit of length, M for the unit of mass, T for the unit of time. These are their definitions:

Unit of length.—The C. G. S. unit of length, or *centimètre*, is equal to the one hundredth part of a mètre. The mètre is the ten millionth part of the quarter of the terrestrial meridian.*

The practical unit of length is represented by copies of the standard mètre deposited at the Observatory of Paris.

Unit of mass.—The C. G. S. unit of mass is the gramme, which is the mass of one cubic centimètre of distilled water at the temperature of 4° C.

The practical unit of mass is represented by the mass of the thousandth part of the standard kilogramme deposited at the Observatory of Paris.

Unit of time.—The C. G. S. unit of time is the second, defined as the $\frac{1}{86,400}$ part of the mean day; this is the only unit which at the present

---

* This is the value which it was intended to have, but owing to errors in the measurement of the quadrant of the meridian it does not represent this value.

day is universally employed; the only one for which we need give no tables of reduction.

1 *minute* = 60 seconds.
1 *hour* = 60 minutes = 36,000 seconds.
1 *day* = 24 hours = 1,440 minutes = 86,400 seconds.

Multiples and sub-multiples.—As the units adopted are found to be sometimes too large, sometimes too small, for practical purposes, multiples and sub-multiples have been established, which represent the decimal multiples, or decimal fractions of these units, and thus the inconvenience of having to write large numbers and large fractions is avoided. These multiples or sub-multiples are indicated by *prefixes*. The following is the nomenclature:

MULTIPLES	*Mega* or *meg* represents	1,000,000	units.
	Myria	10,000	,,
	Kilo	1,000	,,
	Hecto	100	,,
	Deca	10	,,
UNITY.			
SUB-MULTIPLES.	*Deci* represents	$\frac{1}{10}$	of a unit.
	Centi	$\frac{1}{100}$	,,
	Milli	$\frac{1}{1,000}$	,,
	Micro or *micr*	$\frac{1}{1,000,000}$	,,

Thus, for example, a megohm represents a million ohms; a milliampère the thousandth part of an ampère, etc.

Decimal notation or exponent.—Instead of using the multiples or sub-multiples, a number is sometimes expressed by considering it as the product of two factors, of which one is a multiple of 10. The exponent of the power of 10 is the characteristic of the decimal logarithm of this number; for fractions the exponent is negative. Thus, for example, the number 459,000,000 would be written 459×10^6 or 45.9×10^7, and the fraction ·0000459 would be written 459×10^{-7} or $·459 \times 10^{-4}$.

The exponent indicates how many places the decimal point must be displaced to the right for positive exponents, and to the left for negative exponents, in order to write the whole number or the fraction in the usual notation.

Dimensions of units.—All the physical units may be deduced from the three fundamental units, of which the symbols are L, M, and T. The relation which unites a derived unit to one or more of the fundamental units is called the dimension of the unit. Thus, for example, a unit of surface is equal to a square described on the unit of length. The unit of volume is equal to a cube the side of each face of which is one unit of length; the respective dimensions of these units are L^2 and L^3. We will give the dimensions of each derived unit in terms of the fundamental units of the C. G. S. system.

Derived units.—There are a great number of derived units. In order to facilitate their examination, we will divide them into five groups: (1) Geometrical units; (2) mechanical units; (3) magnetic units; (4) electrical units; (5) various other units not included in the first four groups, and which are in constant use in the application of electricity.

1. GEOMETRICAL UNITS.

There are three geometrical units, one fundamental unit (that of length) and two derived units (unit of area and unit of volume). In tables farther on (pages 37 to 39) will be found the relations between the geometrical units of the C. G. S. system and the units still in use in different countries.

C. G. S. unit of length.—The centimètre, which has already been defined.

C. G. S. unit of area is the *square centimètre;* it is the area of a square of one centimètre side. In practice use is made according to circumstances of the square millimètre, of the square centimètre, of the square mètre, of the hectare, or of the square kilomètre. The dimensions of the unit of area are $[L^2]$.

C. G. S. unit of volume.—The cubic centimètre; it is the volume of a cube one centimètre in side. In practice, according to circumstances, the cubic millimètre, cubic centimètre, cubic decimètre, or litre and cubic mètre (*see* table, page 34), are used. The dimensions of the unit of volume are $[L^3]$.

Units of length.—Besides the units of length which are to be found in the table on page 33, the following units are sometimes employed:

In *England*:

1 fathom	= 2 yards.
1 furlong	= $\frac{1}{8}$ of a mile = 220 yards.
1 mil.	= $\frac{1}{1,000}$ of an inch = ·00254 cm.

In *France*:

The myriamètre	= 10,000 mètres.
The lieue terrestre	= 4,000 ,, (little used).
The mille marin (nautical mile) or *knot*	= 1,852 ,,

At *sea* they also use

The brasse	= 5 feet	= 1·624 mètres.
Nœud	= $\frac{1}{120}$ of a mille marin	= 15·436 ,,
New encâblure		= 200 ,,

Each knot of the log-line run out in the 30 seconds of the sand-glass, or in 120th part of an hour, corresponds to the speed of one *mille marin* per hour. Thus, 9 knots run out in 30 seconds indicate a speed of 9 milles or 3 lieues marines per hour.

In *microscopy*, the *micron* (plural *micra*) is used, which is equal to one-millionth of a mètre.

In *Germany* the metric system has been official since the 1st January, 1872.

The millimètre is called	Strich.
The centimètre	Neuzoll.
The mètre	Stab.
The decimètre	Ke'te.

In *Austria* the metric system has been compulsory since 1876.

In *Russia* the unit is the *sagene*, equal to 2·13356143 mètres. The sagene is 3 archines, 7 feet, 48 verschocks, 84 duimes (or inches), 840 linia (lines); a *viersta* (verst) = 500 sagenes = 1066·78 mètres.

Units of area and volume.

In *Germany*:

The square mètre is called	Quadratstab.
The cubic mètre ,,	Kubikstab.
The hectolitre ,,	Fass.
The litre ,,	Kanne.

In *England* the square mile = 2·59 square kilomètres, or one square kilomètre = ·386 square mile.

FRENCH AND ENGLISH UNITS OF LENGTH.

NAME OF UNIT.	CENTI-MÈTRE.	MÈTRE.	KILO-MÈTRE.	INCH.	FOOT.	YARD.	STATUTE MILE.	NAUTICAL MILE.
Centimètre	1	·01	—	·3937	—	—	—	—
Mètre	100	1	·001	39·37	3·281	1·093633	—	—
Kilomètre	100,000	1000	1	—	3280·899	1093·633	·62138	·54
Inch or pouce . .	2·5399	—	—	1	·0833	·0278	—	—
Foot or pied . . .	30·4797	·3048	—	12	1	·3333	—	—
Yard	91·4392	·9144	—	36	3	1	—	·49285
Statute mile . . .	—	1609·31	1·609	—	5280	1760	1	·867422
Nautical mile, knot, or nœud . . .	—	1852·3	1·852	—	6087	2029	1·15284	1

FRENCH AND ENGLISH UNITS OF VOLUME AND CAPACITY.

NAME OF UNIT.	CUBIC CENTI- MÈTRE.	CUBIC DECI- MÈTRE, or litre.	CUBIC MÈTRE.	CUBIC INCH.	CUBIC FOOT.	CUBIC YARD.	PINT.	GALLON.
CUBIC CENTIMÈTRE	1	·001	—	—	—	—	—	—
Cubic decimètre, or litre	1,000	1	·001	61·02705	·035317	—	1·76077	—
Cubic mètre	1,000,000	1000	1	—	35·317	1·3080	1760·77	220·0967
Cubic inch	16·38618	·016386	—	1	—	—	·0291	—
Cubic foot	—	28·316	—	1728	1	·0370	—	—
Cubic yard	—	764·535	·76433	46656	27	1	—	—
Pint	—	·56793	—	34·66	—	—	1	·125
Gallon	—	4·54346	—	277·274	6·232106	—	8	1

UNITS OF AREA.

NAME OF UNIT.	SQUARE CENTIMÈTRE.	SQUARE MÈTRE.	SQUARE INCH.	SQUARE FOOT.
Square centimètre . .	1	·0001	·15501	·001764
Square mètre	10,000	1	1550·1	10 764
Square inch	6·4516	·00064516	1	·00694
Square foot	929·01	·0929	144	1
Square yard	8361·09	·8361	1296	9

2. MECHANICAL UNITS.

Those of velocity, acceleration, force, and work. These are their definitions:

The **C. G. S. unit of velocity.**—That of a body moving in a straight line with a uniform motion, and passing over one centimètre in one second. Its dimensions are $\left[\dfrac{L}{T}\right]$ or $[L\,T^{-1}]$. In practice velocity is expressed according to circumstances, in mètres per second, in mètres per minute, or in kilomètres per hour, or in feet per second, or miles per hour.

The **C. G. S. unit of acceleration.**—That of a body of which the velocity increases one centimètre per second. The acceleration of a body falling freely in vacuo under the action of gravity is represented by g.

Its value varies with the latitude of the place, and the height of the point above the level of the sea. The relation between the value of g and the length l of the seconds pendulum is $g = \pi^2 l$.

The table on page 36 gives the values of l and g in centimètres for the most important places on the globe. The difference between the greatest and least value of g is $\dfrac{1}{196}$ of the mean value. In all tables we have taken $g = 981$ centimètres. The dimensions of the unit of acceleration are $\left[\dfrac{L}{T^2}\right]$ or $[L\,T^{-2}]$.

Values of the Acceleration due to Gravity and the Length of the Seconds Pendulum (*Everett*).

	Latitude.	g.	l.
Equator	0° 0′	978·1	99·103
Latitude 45°	45 0	980·61	99·356
Munich	48 9	980·88	99·384
Paris	48 50	980·94	99·39
Greenwich	51 29	981·17	99·413
Göttingen	51 32	981·17	99·414
Berlin	52 30	981·25	99·422
Dublin	53 21	981·32	99·429
Manchester	53 29	981·34	99·430
Belfast	54 36	981·43	99·440
Edinburgh	55 37	981·54	99·451
Aberdeen	57 9	981·64	99·466
Pole	90 0	983·11	99·61

g and *l* are given in centimètres. To reduce this table to feet divide by 30·4797.

The **C. G. S. unit of force** is called a *dyne*. It is the force which, acting on a mass of 1 gramme for 1 second, gives it a velocity of 1 centimètre per second. The dimensions of the unit of force are $\left[\dfrac{ML}{T^2}\right]$ or $[MLT^{-2}]$.

The C. G. S. unit of force, or dyne, which is indispensable for the establishment of electrical units, is not yet employed in practice; forces are generally expressed as in terms of *weights*. It is thus important to establish the relations which unite the C. G. S. unit of force, or *dyne*, to the practical unit, or *weight of the gramme*. When a body falls in vacuo gravity imparts to it at the end of one second a velocity equal to *g* centimètres per second. As forces are proportional to accelerations, the force which acts on a unit of mass under the influence of the earth's gravity is therefore *g* dynes. The force which acts upon the mass of one gramme, that is to say, the weight of one gramme being *g* dynes, the dyne is therefore equal to $\dfrac{1}{g}$ gramme. Adopting the number 981 as the value of *g*, which corresponds to middle latitudes (about 50°) the weight of a gramme is equal to 981 dynes, and the dyne is equal to $\dfrac{1}{981}$ of the weight of one gramme. The weight of a gramme varies with the latitude, whilst the mass of a gramme is a constant quantity.

Unit of weight.—The practical unit is the weight of a gramme or the force necessary to support in vacuo 1 cubic centimètre of distilled water at a temperature of 4° C. The table on page 38 shows the relation between the different units of force and weight employed in France and in England, taking 981 as the value of g.

In France they still use sometimes

The *carat* for diamonds	= 205·5	milligrammes.
The *livre*	= 500	grammes.
The *quintal metrique*	= 50	kilogrammes.
The *tonne* or *tonneau*	= 1000	kilogrammes.

In Germany the unit of weight is the kilogramme = 2 *livres* (2 *pfunds*).

The décagramme is called *neuloth*, 50 kilogrammes *centner*, and the half kilogramme *pfund*; 1,000 kilogrammes *tonne*.

In Russia the unit is the commercial pound (1 *founte* = 409·511663 grammes); it contains

16 onces, 32 loths, 96 zolotnicks, 9,216 dolis.
Poude = 40 livres.
Berkowetz = 10 poudes = 400 livres.
Sea-ton = 6 Berkowetz = 982·5 kilogrammes.

The **C. G. S. unit of work or energy** is called the *erg*; it is the work produced by a force of 1 dyne acting over a distance of 1 centimètre; it may be called the centimètre-dyne. It has not yet been employed in practice, but gramme-centimètres, grammètres, kilogrammètres, and foot-pounds are used. The kilogrammètre is the work produced by a weight of 1 kilogramme falling from a height of 1 mètre. The dimensions of a unit of work are $\left[\dfrac{ML^2}{T^2}\right]$ or $[ML^2\, T^{-2}]$.

The tables on pages 38, 39, show the relations between the different units of work used in practice.

In gunnery sometimes the ton mètre (which is equal to 1,000 kilogrammètres) or the foot-ton is used.

English horse-power.—The horse-power is equal to 550 foot-pounds per second, 33,000 per minute, 1,980,000 per hour.

French cheval-vapeur.—The French unit of work produced or expanded by engines per unit of time. It is equal to 75 kilogrammètres per second, or 542·4825 foot-pounds per second. We may say indifferently that an engine produces 10 chevaux-vapeur, or 750 kilogrammètres, per second.

UNITS OF WEIGHT AND FORCE ($g = 981$ cm. $= 32{\cdot}195$ feet).

NAME OF UNIT.	MILLI-GRAMME.	GRAMME.	KILO-GRAMME.	DYNE.	MÉGADYNE.	GRAIN.	OUNCE (avoirdu-pois).	POUND (avoirdu-pois).
French.								
Milligramme		0·001	—	0·981	—	·01543235	—	—
GRAMME	1,000	1	0·001	981	—	15·43235	—	—
Kilogramme	1,000,000	1,000	1	981000	0·981	15432·35	—	2·204621
Absolute C.G.S.								
DYNE (C. G. S. unit)	1·01937	·001019	—	1	—	0·0310666	—	—
Megadyne	1,019,370	1019·37	1·01937	1,000,000	1	—	—	—
English.								
Grain (troy)	64·799	0·06480	—	63·57	—	1	—	·000142857
Ounce (avoirdupois)	—	28·3495	—	27,800	—	437·31	1	9·0625
Pound (avoirdupois) or (lbs.)	—	453·59	0·45359	445,000	49·8	7000	13	1
Cwt. (hundred-weights) (quintal)	—	50802	50·80258	—	—	—	1792	112
Ton	—	—	1016·05	—	997	—	35840	2240

UNITS OF WORK ($g = 981$ cm. $= 32\cdot195$ feet).

NAME OF UNIT.	GRAMME CENTI-MÈTRE.	GRAM-MÈTRE.	KILO-GRAM-MÈTRE.	ERG.	MEG-ERG.	FOOT-GRAIN.	FOOT-POUND.	FOOT-TON.
Gramme-centimètre	1	0·01	—	981	—	—	—	—
Grammètre . .	100	1	0·001	98,100	0·0981	—	0·00723	—
Kilogrammètre . .	100,000	1000	1	—	98·1	—	7·2331	0·00325
Erg	0·00109	—	—	1	0·000001	—	—	—
Meg-erg . . .	1093·67	10·9367	0·0109	1,000,000	1	—	—	—
Foot-grain . .	1·973	—	—	1,937	—	1	—	—
Foot-pound . .	13800	138	0·138	—	13·56	7,000	1	—
Foot-ton . .	—	—	309	—	30400	—	2240	1

These are the respective values of the English horse-power, and the French cheval-vapeur.

1 horse-power	=	75·9 kilogrammètres per second.
,, ...	=	7,460 meg-ergs per second.
,, ...	=	550 foot-pounds per second.
,, ...	=	1·0139 *cheval-vapeur*.
1 *cheval-vapeur*	=	75 kilogrammètres per second.
,, ...	=	7,360 meg-ergs per second.
,, ...	=	542·4825 foot-pounds per second.
,, ...	=	·9863 horse-power.

Hour horse-power and cheval heure.—Units employed in testing accumulators. These are true units of work (the horse-power and cheval-vapeur are units of activity or *rate of doing work*). The nomenclature is bad, but is too firmly established to be at once abolished. They represent the quantity of energy given by a horse-power, or cheval-vapeur, in one hour.

1 hour horse-power	=	75·9 × 3,600 kilogrammètres.
,, ...	=	7,460 × 3,600 meg-ergs.
,, ...	=	550 × 3,600 foot-pounds.
,, ...	=	1·0139 cheval heure.
1 cheval heure	=	75 × 3,600 kilogrammètres.
,, ...	=	7,360 × 3,600 meg-ergs.
,, ...	=	542·4825 × 3,600 foot-pounds.
,, ...	=	·9863 hour horse-power.

Watt- or volt-ampère.—A unit proposed to the British Association in 1882 by Sir William Siemens, and employed in some recent works. It is the activity or rate of doing work of one ampère multiplied by one volt.

1 *watt*	=	$\frac{1}{9\cdot81}$ kilogrammètre per second.
1 *horse-power*	=	746 watts.
1 *cheval vapeur*	=	736 watts.

3. MAGNETIC UNITS.

These units are not yet very much employed in practice. Their purpose is principally to serve as links between the geometrical and mechanical units, and the electro-magnetic units which we will define a little later on.

Unit magnetic pole.—A magnetic pole, whose intensity is equal to 1 C. G. S. unit, is that which repels a similar pole placed at a distance of 1 centimètre, with a force of 1 dyne. It has no special name; its dimensions are $\left[M^{\frac{1}{2}} L^{\frac{3}{2}} T^{-1} \right]$.

Unit of intensity of a magnetic field.—The intensity of a magnetic field is measured by the force which it exerts on a magnetic pole of one unit intensity. The intensity of a magnetic field is equal to one C. G. S. unit, when the force which acts on a unit magnetic pole in this field is equal to one dyne. Its dimensions are $\left[M^{\frac{1}{2}}L^{-\frac{1}{2}}T^{-1}\right]$.

ELECTRICAL UNITS.

There are two systems of electrical units derived from the fundamental C. G. S. units. The first, called the *electrostatic* system, based upon the forces exerted between two quantities of electricity. The second, called the *electro-magnetic*, based on the forces exerted between two magnetic poles. The first has a purely scientific interest, whilst the second has formed the basis for the electrical units adopted by the Congress of Electricians held at Paris in 1871. It is the one used in this work.

4. ELECTRO-MAGNETIC UNITS.

C. G. S. units.—Practical units.—There are five electro-magnetic units in the C. G. S. system; they are deduced from the fundamental, geometrical, mechanical, and magnetic units, by the definitions which we are about to point out; but as their employment would lead to the use of too large or too small numbers, units have been adopted in practice, which are decimal multiples, or sub-multiples of the C. G. S. units, and to avoid confusion, these practical units have now special names to distinguish them from the C. G. S. units. The following table shows the relations between the C. G. S. units and the corresponding practical units, the symbols which represent them, and the dimensions of each unit in terms of the fundamental units:

TABLE OF ELECTRO-MAGNETIC UNITS.

QUANTITIES.	SYMBOL.	NAME OF PRACTICAL UNIT.	NUMBER OF C.G.S. UNITS IN ONE PRACTICAL UNIT.	DIMENSIONS OF UNIT.
Resistance	R	*Ohm.*	10^9	LT^{-1}
Electromotive force	E	*Volt.*	10^8	$M^{\frac{1}{2}}L^{\frac{3}{2}}T^{-2}$
Current strength	C	*Ampère.*	10^{-1}	$M^{\frac{1}{2}}L^{\frac{1}{2}}T^{-1}$
Quantity	Q	*Coulomb.*	10^{-1}	$M^{\frac{1}{2}}L^{\frac{1}{2}}$
Capacity	C	*Farad.*	10^{-9}	$L^{-1}T^2$

Before giving the relative values of the different electrical units employed by electricians, we will define the C. G. S. units which are now used, as well as the practical units which have been deduced from them.

Unit of current strength.—A current has a strength equal to one C. G. S. unit if, when it passes through a circuit one centimètre long, bent into the form of an arc of a circle of one centimetre radius, it exerts a force of one dyne on a unit magnet pole placed at the centre of the circle. The practical unit of current strength is called the ampère, and is equal to 10^{-1} C. G. S. units. (It is the old *Weber per second* often called the *Weber*.)

Unit of quantity.—The C. G. S. unit is the quantity of electricity which passes through a circuit in one second when the strength of the current is equal to one C. G. S. unit. The practical unit of quantity is called a *coulomb*, and is equal to 10^{-1} C. G. S. units. (*This is the old Weber*.)

Unit of electromotive force.—When a certain quantity Q of electricity traverses a conductor under the influence of an electromotive force E, the work produced is equal to the product QE. This being so, the C. G. S. unit of electromotive force is that which is necessary in order that a unit of quantity may develop one C. G. S. unit of work, or one erg. The practical unit of electromotive force is called the *volt*; it is equal to 10^8 C. G. S. units.

Unit of resistance.—A conductor has a resistance equal to one C. G. S. unit when unit e. m. f. (or, more correctly, unit difference of potential) between its two ends causes a unit of current to pass through it.

The practical unit of resistance is called the ohm; it is equal to 10^9 C. G. S. units. Ohm's law $C = \dfrac{E}{R}$ establishes the relation between these three practical units of current strength, e. m. f., and resistance, which may be written

$$1 \; ampère = \frac{1 \; volt}{1 \; ohm}.$$

Unit of capacity.—A condenser has a capacity equal to one C. G. S. unit, if, when charged to a potential of one C. G. S. unit, it contains a quantity of electricity equal to one C. G. S. unit. The practical unit of capacity is called a *farad*; it is equal to 10^{-9} C. G. S. units. As even the farad is too large a quantity for practical purposes, the micro-farad is most commonly used, of which the value is equal to 10^{-15} C. G. S. unit, or

10^{-6} farad. A condenser of one micro-farad charged to the potential of one volt contains a quantity of electricity equal to one micro-coulomb. Before the labours of the British Association, and the sanction given to them by the International Congress of Electricians, which will have the effect of soon rendering the employment of these practical units which we have just described universal, many physicists, working upon bases not so well co-ordinated as those of the British Association, employed certain units which are falling out of use, but which may still often be found in many original treatises and memoirs. We will bring the most important of these units together in the form of synoptic tables, giving to each of them their values in terms of the units of the Congress of 1881.

COMPARISON OF ELECTRICAL UNITS EMPLOYED BY DIFFERENT PHYSICISTS.

Units of resistance.— The practical unit adopted by the Congress is called the *ohm*, and is equal to 10^9 C. G. S. units: that is its theoretical value. In practice, the ohm is represented by standards constructed by the British Association in 1864, and of which the value is about 10^9 C. G. S. units. Numerous verifications of the standard established by the British Association have shown that the difference between the real value of the standard and its theoretical value is about 1 per cent.*

To obtain the value of the ohm more exactly, the Congress has decided that the practical unit of resistance should be represented by a column of mercury of one square millimètre section, at a temperature of $0°$ C., and that an International Commission should be appointed to determine exactly the length of this column of mercury. The Conference which met at Paris in the month of October, 1882, was adjourned to the month of October, 1883, to definitely establish the true value of the ohm. We must, for the present, accept the ohm of the British Association, and it is this value to which we reduce that of all the other units. It will be easy, later on, to reduce these figures to the value of the ohm of the Congress by multiplying them by a coefficient.

The next table (page 45) gives the relative values of the units of resistance used by different physicists.

1 and 2. *English units*, now abandoned, based on the English foot and second.

---

* The most recent experiments made by Lord Rayleigh and Mrs. H. Sidgwick ("Proceedings of the Royal Society," of April, 1882) have given:

Then,
$$1 \text{ ohm or B. A. unit} = \cdot 98651 \times 10^9 \text{ C. G. S. units.}$$
$$\text{One Siemens unit} = \cdot 94130 \times 10^9 \text{ C. G. S. units.}$$

3. *Jacobi's unit.*—The resistance of a certain copper wire 25 feet long, and weighing 345 grains.

4. *Weber's absolute unit*, based on the mètre and the second.

5. *Siemens' unit*, represented by the resistance of a column of mercury one mètre long and of a square millimètre section at 0° C.

6. British Association unit, or *ohm*.

7, 8, and 9. Units constructed by *Digney*, *Breguet*, and the *Administration Suisse*, represented by the resistance of an iron wire one kilomètre long and four millimètres diameter, at an unknown temperature.

10. *Matthiessen's unit.*—A standard mile of pure annealed copper wire $\frac{1}{16}$ inch diameter, at a temperature of 15·5° C.

11. *Varley's unit.*—A standard mile of a special copper wire $\frac{1}{16}$ inch in diameter.

12. One *German mile* (8,238 yards) of iron wire $\frac{1}{6}$ inch diameter.

Out of these twelve units of resistance, only the ohm and the Siemens unit are now used. In France the kilomètre of telegraphic wire also tends to disappear, and give place to the ohm.

Unit of electromotive force.—There is no standard of e. m. f. in existence which gives exactly one volt. Experimenters often express electromotive forces by taking the battery which they use as a standard. Amongst the standards which are most commonly used are:

The Daniell element, which, put up under certain conditions, has an e. m. f. of 1·079 volts.

The Bunsen element can only be used for rough measurements.

Latimer Clark's standard cell is very constant when it is on open circuit; its e. m. f. is 1·457 volts.

Gaugain's thermo-electric unit, represented by the symbol $\frac{Bi-Cu}{0°-100°}$, is the e. m. f. of a thermo-electric bismuth-copper element of which the junctions are maintained at the temperatures of 0° and 100°. It is no longer employed. Its value also is very small, $\frac{1}{197}$ of a Daniell, or $\frac{1}{182·6}$ of a volt.

Units of current strength.—The greater part of practical units of current strength are based on electrolytic actions; those which have been most used up to the present time are:

Jacobi's unit.—A continuous and constant current producing one cubic centimètre of mixed gas per second in a voltmeter at temperature 0° C., and pressure 760 millimètres of mercury.

TABLE OF DIFFERENT UNITS OF ELECTRICAL RESISTANCE.

(Report of the Committee on Electrical Standards.)

RESISTANCE.

Numbers.	Name of Unit.	1. Absolute $\frac{foot}{second} \times 10^7$	2. Thomson's Old Unit.	3. Jacobi.	4. Weber's Absolute $\frac{mètre}{second} \times 10^7$	5. Siemens.	6. B.A. Unit or Ohm.	7. Digney.	8. Breguet.	9. Swiss.	10. Matthiessen.	11. Varley.	12. German Mile.
1.	Absolute $\frac{foot}{second} \times 10^7$	1·00	·9520	·4788	·3316	·3196	·3048	·03260	·03123	·02924	·02243	·01190	·005307
2.	Thomson's unit	1·0505	1	·5029	·3483	·3358	·3202	·03455	·03279	·03071	·02337	·01251	·005574
3.	Jacobi	2·088	1·983	1	·6925	·6674	·6367	·03869	·052	·05106	·04686	·02496	·01108
4.	Weber's absolute $\frac{mètre}{second} \times 10^7$	3·015	2·871	1·444	1	·9635	·9191	·09919	·09416	·08817	·06767	·03591	·01655
5.	Siemens	3·129	2·979	1·498	1·038	1	·9537	·1030	·0975	·0915	·07027	·03725	·01661
6.	B.A. unit or ohm	3·281	3·123	1·57	1·083	1·0486	1	·1079	·1024	·0959	·0733	·03905	·01741
7.	Digney	30·4	28·94	14·56	10·08	9·71	9·266	1	·9491	·8889	·6822	·3620	·1613
8.	Breguet	32·03	30·5	15·34	10·62	10·23	9·769	1·054	1	·9365	·7187	·3814	·17
9.	Swiss	34·21	32·56	16·38	11·34	10·93	10·42	1·125	1·068	1	·7675	·4072	·1815
10.	Matthiessen	44·57	42·43	21·34	14·78	14·23	13·59	1·66	1·391	1·303	1	·5303	·2365
11.	Varley	84·01	79·96	40·21	27·85	26·83	25·61	2·763	2·622	2·456	1·885	1	·4457
12.	German mile	188·4	179·4	90·22	62·48	60·29	57·44	6·198	5·882	5·509	4·228	2·243	1

UNITS OF MEASUREMENT.

The unit, $\frac{\text{Daniell}}{\text{Siemens unit}}$, indicated in Germany by the symbol $\frac{\text{Daniell}}{\text{U. S.}}$, is the strength of the current produced by a Daniell element in a circuit, the total resistance of which is one Siemens unit; this current deposits 1·38 grammes of copper per hour.

Atomic current.—A unit employed in Germany. The strength of a current which, passing through a voltmeter for twenty-four hours, or 86,400 seconds, disengages one gramme of hydrogen.

In telegraphy the *milliatom* has been used, comparable to the milliampère, but the milliampère is beginning to be substituted everywhere for the milliatom.

Various units.—We often find in memoirs the strength of currents expressed by the weight or volume of gas disengaged during a given time. It is easy to reduce these strengths to ampères by remembering that a current of one ampère disengages ·172 cubic centimètre of mixed gas per second at 0° C. and pressure 760 mm.

The weight of water decomposed is 92 microgrammes.* The weight of hydrogen produced is 10·4 microgrammes, and the weight of oxygen 81·6 microgrammes.

TABLE OF UNITS OF CURRENT STRENGTH.

NAME OF UNIT.	AMPÈRE.	JACOBI.	$\frac{\text{DANIELL}}{\text{U.S.}}$	ATOMIC CURRENT.
Ampère	1	10·32	·862	·90009
Jacobi	·0961	1	·08283	0·86198
$\left(\frac{\text{Daniell}}{\text{U.S.}}\right)$	1·16	12	1	1·0441
Atomic current	1·111	11·5	·95775	1

Unit of quantity.—The unit of quantity which is at present used is the *coulomb*. This is the quantity of electricity which passes through a circuit in one second, when the current strength is one ampère. Sometimes, however, the quantity of electricity is expressed in terms of the weight of metal deposited by a current in a decomposition cell, or by the volume of gas disengaged in a voltmeter. The table of electrochemical equivalents (*see* Fourth Part) enables us to reduce to coulombs the quantity of electricity expressed by weights of deposited metals. When the quantities of electricity are expressed by volumes of gas, the value in coulombs is found by remembering that one coulomb of electricity passing

---
* Some authors give the figures 93·78 microgrammes.

ELECTROSTATIC UNITS.   47

through a voltmeter produces ·172 cubic centimètre of mixed gas at the temperature 0° C., and pressure 760 mm. The weight of water decomposed by one coulomb is 92 microgrammes, and the weight of the hydrogen evolved is 10·4 microgrammes.

Ampère hour, or hour ampère.—The quantity of electricity which passes through a circuit in one hour when the current strength is one ampère.

One hour ampère equals 3,600 coulombs. (Used in testing accumulators.)

Unit of capacity.—The unit most generally employed for condensers and submarine cables is the microfarad, which we have already defined.

Electrical units of Messrs. Siemens of Berlin.—Since the Paris Congress of 1881 all the measuring apparatus of this firm has been based on the following units :

One ohm = 1·0615 Siemens units.

The standard ohm of the British Association, or B. A. unit, = ·9935 ohm.*

The standard ohm of the British Association, or B. A. unit, = 1·0493 Siemens units.

One hour ampère deposits 3·96 grammes of silver.

One coulomb deposits ·0010833 gramme of silver.

ELECTROSTATIC UNITS.

Although the electrostatic units are very rarely used in practice, it is useful to define them here.

Electrostatic unit of quantity.—The unit of quantity is that which, placed at a distance of one centimètre from a similar equal quantity, repels it with the force of one dyne. Dimensions $[M^{\frac{1}{2}} L^{\frac{3}{2}} T^{-1}]$.

Electrostatic unit of difference of potential.—There is unit difference of potential between two points when it requires one unit of work, or one erg, in order to move a quantity of electricity equal to one unit from the one point to the other. Dimensions $[M^{\frac{1}{2}} L^{\frac{1}{2}} T^{-1}]$.

---

i.e. the ohm defined by theory as equal to 10^9 C. G. S. units.

Unit of electrostatic capacity.—The capacity of a conductor is one unit when one unit quantity of electricity raises its potential one unit. Dimensions [L].

A sphere of one centimètre radius has a capacity equal to one unit. The capacities of spheres are proportional to their radii.

Relation between electrostatic and electro-magnetic units.—The ratio between the electrostatic unit of quantity and the electro-magnetic unit of quantity, has for its dimensions $\left[\frac{L}{T}\right]$. This expression is equivalent to a velocity, and is represented by the letter v; the numerical value of v varies between $2 \cdot 825 \times 10^{10}$, and $3 \cdot 1074 \times 10^{10}$ centimètres per second. The value now adopted is that given by Professors Ayrton and Perry:

$$v = 2 \cdot 98 \times 10^9 \text{ centimètres per second.}$$

This velocity is the same as that found for the velocity of light.

5. VARIOUS UNITS.

Under this general title we will discuss the physical units, which, although they do not belong to the C. G. S. system, and are not officially recognised by the International Congress of Electricians, are nevertheless accepted almost universally by scientific men, and are to a certain extent directly derived from the C. G. S. system.

Units of pressure.—The unit of pressure in the C. G. S. system would be equal to a unit of force exerted on a unit of area; that is to say, would be equal to one dyne per square centimètre. This unit has only a theoretical value, and is not employed in practice.*

In France they reckon by atmospheres, and by kilogrammes per square centimètre. The atmosphere, or atmospheric pressure, is equal to the pressure of a column of mercury, 760 millimètres in height, at 0° C.; or to a column of water 10·33 mètres in height. In England the atmosphere is thirty inches of mercury. The kilogramme per square centimètre is equal to the pressure of a column of water 10 mètres high. As these two units do not differ greatly one may be used for the other without introducing any very serious error. In England pressure is generally reckoned in pounds per square foot, and pounds per square yard. The table on page 50 shows the relations between these different units.

---

* Some English physicists use it in scientific work. (See Everett, *Units and Physical Constants.*)

Unit of temperature.—The unit of temperature generally adopted is the degree centigrade or degree Celsius (C. by abbreviation). It is founded on the thermal properties of distilled water at the atmospheric pressure (760 millimètres). In the practical centigrade thermometric scale, the 0° is the temperature of melting ice; 100° that of water boiling under a pressure of 760 millimètres of mercury; and the centigrade degree is the one-hundredth part of the difference of temperature between 0° and 100°. In *Réaumur's* scale 0° corresponds to melting ice, and 80° to the temperature of boiling water. In Fahrenheit's scale the temperature of melting ice is represented by 32°, and that of boiling water by 212°. The following table shows the relation between the degrees of these three scales.

RELATION BETWEEN THE DIFFERENT THERMOMETRIC DEGREES.

Thermometer Scale.	Centigrade.	Réaumur.	Fahrenheit.
Centigrade or Celsius	1	·8	1·8
Réaumur	1·25	1	2·25
Fahrenheit	·55556	·4444	1

Temperatures are sometimes recorded in a scale which is called that of *absolute temperature*. The value of the degree is the same as that of the degree centigrade, but the 0 of the absolute scale corresponds to −273° of the centigrade thermometer. To reduce temperatures expressed in the absolute scale to the centigrade scale, subtract 273 from the number which expresses the absolute temperature.

Units of heat.—The practical unit of heat used in France, and generally adopted by scientific writers, is called **the calorie**, and is the quantity necessary **to raise** one kilogramme of water 1° C. The theoretical unit of heat is, as yet, rather badly defined, because the specific heat of water varies with its temperature, and the temperature adopted as the standard varies with different authorities. Generally the temperature is selected between 0° and 4° C. Some physicists have adopted a unit a thousand times smaller than this, *i.e.* a quantity of heat necessary to raise one gramme of water one degree centigrade; unfortunately they have also called this unit by the name of calorie, because it is more directly derived from the C. G. S. system, as it depends upon the gramme as a unit of mass.

E

UNITS OF PRESSURE ($g = 981$ cm. $= 32\cdot 195$ feet).

NAME OF UNIT.	ATMO-SPHERE.	KILO-GRAMME PER SQUARE MÈTRE.	KILO-GRAMME PER SQUARE CENTIMÈTRE.	DYNE PER SQUARE CENTIMÈTRE.	POUND PER SQUARE FOOT.	POUND PER SQUARE INCH.
Atmosphere (76 mm. of mercury at 0°)	1	10,330	1·033	1,014,000	2,118	14·67
Kilogramme per square mètre	—	1	0·0001	98·1	·205	·0142
Kilogramme per square centimètre	0·968	10,000	—	981,000	2,050	14·2
Dyne per square centimètre	—	—	—	1	·00211	—
Pound per square foot	0·00047	4·88697	—	479	1	0·0067
Pound per square inch	0·0681	703·876	0·0704	69,000	144	1

Pressure of 30 inches of mercury at 0° C. . . . = 1,016,300 dynes per square centimètre.

Pressure of 1 inch of mercury at 0° C. . . . = 33,880 ,, ,, ,,

UNITS OF ENERGY, HEAT, AND WORK (g = 981 cm. = 32·195 feet).

NAME OF UNIT.	CALORIE (g.-d. C.)	CALORIE (kg.-d. C.)	MEG-ERG.	KILO-GRAM-MÈTRE.	POUND-DEGREE C.	POUND-DEGREE FAHREN-HEIT.	FOOT-POUND.	WATT.
Calorie (g.-d. C.) .	1	0·001	41·6	0·424	0·0022	0·004	3·0668	4·16
Calorie (kg.-d. C.) .	1,000	1	41,600	424	2·2056	3·968	3,066·83	4,160
Meg-erg	0·00243	—	1	0·0102	—	—	·0737	·1
Kilogrammètre . .	2·378	0·00236	93·1	1	0·00515	0·00926	7·23	9·31
Pound-degree C. .	—	0·4545	19,100	194	1	0·5556	1403·22	1,910
Pound-degree F.-h.-renheit (*Thermal unit*) . . .	—	0·252	10,600	108	1·8	1	780·2	1,060
Watt	·243	—	10	·102	—	—	·7377	1
Foot-pound . . .	·3262	·00032	13·55	·138	·000712	·00128	1	1·355

In order to avoid confusion in this work, we will always distinguish the calorie (kilogramme degree) from the calorie (gramme degree). This last unit is also sometimes called the milli-calorie.

In England they use the *pound-degree-centigrade*, or *pound-degree-Fahrenheit*, or *thermal unit*.

The *pound-degree-centigrade* is a bastard unit based upon the English pound and the centigrade degree. Its name defines it sufficiently.

The *pound-degree-Fahrenheit* or *thermal unit* is the quantity of heat necessary to raise one English pound of water one degree Fahrenheit. The table on page 51 shows the relations between these two units.

Mechanical equivalent of heat.— The value generally adopted for the mechanical equivalent of heat is the following:

1 *calorie* (kg.-d) = 424 kgm. = 3065·83 foot-pounds = 5·57 h. p.

When energy is considered under its different forms of work, heat, or electricity, it is expressed, according to circumstances, in units of work or units of heat. The table on page 51 shows the value of the relations between the different units of energy most commonly used, calorie, meg-erg, kilogrammètre, and foot-pound.

PHOTOMETRIC UNITS.

ENGLAND.—The unit is the candle (or *Parliamentary standard*) of spermaceti, a candle $\frac{7}{8}$ inch in diameter burning 120 grains per hour. (This standard sometimes varies 30 per cent.)

FRANCE.—The unit is the *Carcel burner*, burning 42 grammes of pure colza oil per hour with a flame of 40 millimètres, under the conditions established by Messrs. J. B. Dumas and Regnault, who devised this system for the purpose of testing the illuminating power of the Paris gas.

One carcel burner = 9·5 candles.

GERMANY.—The standard is a paraffin candle 20 millimètres in diameter burning with a flame 5 centimètres high.

One carcel burner = 7·6 German candles.

VARIOUS UNITS.—*Schwendler* has proposed a strip of platinum, of given dimensions, raised to incandescence by a current of given strength. M. *Violle*, in 1881, proposed as a standard the light emitted by a platinum surface one square centimètre in area kept at its point of fusion. Neither of these units is yet adopted in practice.

ced. Mech. Eng.

Third Part.

METHODS OF MEASUREMENT AND MEASURING INSTRUMENTS.

The applications of electricity are so various, that an electrician's pocket-book ought to comprise all the measuring apparatus and all the methods of measurement of geometrical, mechanical, and physical quantities. Under these conditions, this pocket-book would become a general treatise far exceeding the limits which we have laid down; so we must content ourselves with rapidly running over those methods and instruments which only interest the electrician indirectly, reserving the greater part of our space for the methods and applications which the electrician uses directly. We will here follow the same order which we have adopted in the explanation of the units of measurement.

GEOMETRICAL MEASUREMENTS.

Geometrical measurements which interest the electrician are of the same kind as those which all engineers generally require. In the Fourth Part of this book we give some common formulæ for the measurement of area and volume, as well as a few figures which are in constant use.

We need only describe one small piece of geometrical measuring apparatus devised for the measurement of electrical conductors.

Micrometer gauge.—This apparatus is a small micrometer

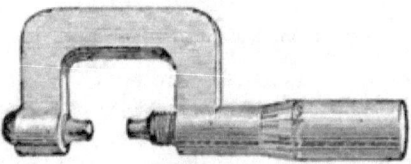

Fig. 7.—Micrometer Gauge.

screw, which is very handy for measuring the diameter of wires. It is sufficiently delicate to measure them in 100ths of a millimètre.

The best patterns have a movable milled head, which "breaks off" as soon as the pressure of the screw exceeds a certain value, so that it is impossible to crush the wire or the plate of which the diameter or the thickness is being measured.

Sometimes round gauges are used, carrying a series of numbered slots, but as these gauges only indicate the number of the wire, and not its diameter, they are falling out of use.

MECHANICAL MEASUREMENTS.

These relate to velocity, force, and work.

Velocity.—The electrician generally has only to measure the velocity of rotation. For this purpose counters (*Sainte, Dechiens*, etc.) are used, which show the mean number of revolutions made by an engine in a given time; or speed indicators (*Buss's* tachymètre, *Murdoch Napier's* show-speed, the velocity indicator of *Marcel Deprez*, etc.), which show the velocity at a given moment directly. The linear velocity v of a point situated at a distance r from the axis is given in mètres per second if r be in mètres, or feet per second if r be in feet, by the formula

$$v = \frac{2\pi r n}{60},$$

n being the number of revolutions per minute.

Measurement of force is either made by comparing the force to be measured to weights, or by means of a traction dynamometer, or by measuring a pressure exerted on a given area.

Measurement of work.—Work being one of the forms of energy, we will point out some of the methods employed for measuring it when we come to consider the measurement of energy.

ELECTRICAL MEASUREMENTS.

Electrical quantities are measured either directly or indirectly. Before describing the methods employed in these measurements, we must describe the instruments which are used. When the instruments have been described, the choice of methods will explain itself, as generally these methods only depend on the apparatus which is at hand and the accuracy which we wish to obtain; so we will rapidly glance over resistance coils, resistance boxes, standards of e. m. f., galvanometers, electrometers, measuring instruments, and accessory apparatus in general use.

RESISTANCE COILS AND RESISTANCE BOXES.

A good resistance box ought to have the following properties: Accuracy of adjustment, small sensibility to variation of temperature, and constancy of resistance under the action of currents or other physical actions. The coils of which the boxes are composed are in general made of German silver, covered with a double coating of silk. The wire is doubled before being coiled, so as to avoid effects of induction, and is wound upon an ebonite or paraffined-wood reel. The English makers use reels made of thin brass, in order that the coils may cool down rapidly after being heated by the passage of a current. When the coil is finished and its resistance corrected, it is steeped in melted paraffin wax, which ensures its insulation, and prevents the effects of moisture causing short-circuiting and so diminishing its resistance. The thickness of the wire of which resistance coils are made does not much matter; it depends only on the nature of the currents which have to pass through the coil, which ought in no case to become sensibly heated. The correction of the coil is easier the thicker the wire is. The resistance of the wire varies with the quantity of nickel in the alloy; the wire ought to be very soft. The coils ought to be measured after they have been coiled, because the necessary tension in coiling them varies the resistance. The usual composition of German silver (maillechort, pacfong, or argentan) is as follows (*V. Regnault*):

Copper	50 parts.
Zinc	30 ,,
Nickel	20 ,,

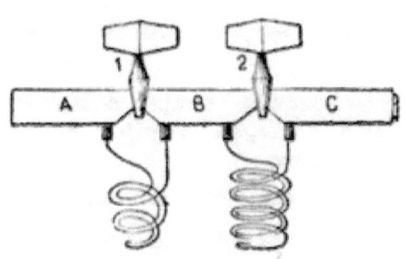 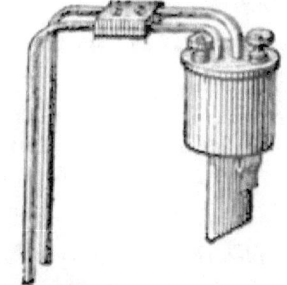

Fig. 8.—Arrangement of Blocks and Plugs.   Fig. 9.—Standard B. A. Unit.

The above diagrams show how the coils are joined to the corresponding blocks, and how the putting in of a plug cuts out the resistance of the corresponding coil by short-circuiting it.

When very great accuracy is required, the coils are made of the silver platinum alloy used for the B. A. standard ohm.

The British Association standard coil.—This represents the ohm. It is formed either of German silver, or of an alloy of 66·6 of silver and 33·4 of platinum. The wire is carefully insulated with two or more coatings, and doubled before being wound, so as to prevent inductive action. The two ends of the wire are soldered to two massive copper rods; the whole coil is steeped in paraffin, and enclosed in a brass case (Fig. 9). By immersing the case in water, the wire may be raised to any temperature desired. These coils are constructed with great care, and are specially employed with the sliding Wheatstone's bridge, or for testing resistance boxes.

Subdivision of the ohm.—A German silver wire (No. 16 B. W. G.), 1·65 millimètres in diameter and of one ohm resistance, is taken. It is then carefully divided into ten equal parts, and at each division a copper wire is soldered, which enable tenths of an ohm to be used. For hundredths of an ohm, a brass wire $\frac{1}{10}$ of an ohm of No. 18 gauge, may be taken and treated in the same way; No. 10 brass wire for thousandths of an ohm. Leaving out of the question considerations of weight, volume, and price, it is better to take the largest possible wire, by which means the greatest accuracy is insured.

Resistance boxes.—Generally coils are chosen of such resistance, that, by their combination, all resistances from 1 to 10,000 ohms can be found. With a minimum number of coils, generally the following numbers are taken:

1 2 2 5 10 10 20 50 100 100 200 500 1000 1000 2000 5000

i.e. sixteen coils in all.

Another combination of coils, which permits of rapid manipulation, enables us to read the resistances at a glance, and to adjust them with very little movement of the plugs, is as follows:

1 2 3 4 10 20 30 40 100 200 300 400 1000 2000 3000 4000

This also requires sixteen coils. In the form of dial resistance box, which Elliott makes, the great consideration has been to have the smallest possible number of plugs. (*See* page 58.)

Bridge box: Post-Office pattern.—This box contains, as the following diagram shows, a series of resistances, enabling us to obtain from 1 to 10,000 ohms resistance; a Wheatstone's bridge, with arms 10, 100, and 1,000, and two contact keys, one for the battery, and the other for the galvanometer; and two plugs marked "infinity," the

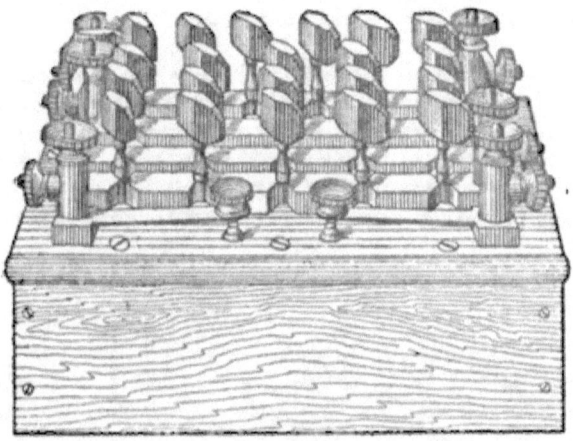

Fig. 10.—Post-Office Bridge Box.

removal of which completely cuts the connections at the points where they are inserted. One of these plugs enables the bridge box to be turned into an ordinary resistance box; the other enables us to insert an

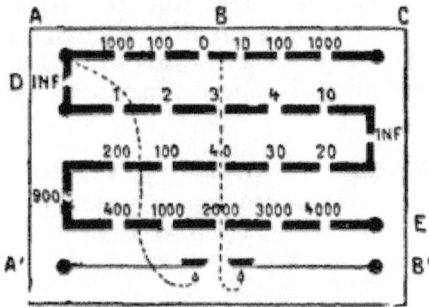

Fig. 11.—Diagram of Connections in the Post-Office Bridge Box.

infinite resistance in one of the branches of the bridge. The most complete pattern has a reversing key, which changes the direction of the battery current without disturbing the connections.

Bridge for cable testing.—This box has no contact keys such as are found in the Post-Office pattern. The contacts are made by means of special accessory apparatus (reversing keys, or tappers, or Morse keys).

Each of the copper blocks which join the coils has a hole in it, in which the plug can be placed when out of use.

Dial box.—This is a bridge box in which the arms have coils of 10, 100, 1,000, and 10,000 ohms. The resistance box proper has five dials, each one formed of a disc of brass, surrounded by a ring divided into ten segments; each segment is connected to its neighbour by a resistance coil. All the coils of any one dial have the same value, 1, 10, 100, 1,000 ohms; the current goes from the segments to the centre

Fig. 12.—Dial Resistance Box.

of the dial. A plug is placed between the centre of the disc and one of the segments. If the plug is at zero, there is no resistance introduced; if it is put at 1, 2, 3, the current has to pass through 1, 2, 3 coils on its way from zero to the centre. The resistance introduced is equal to the sum of the resistance plugged up. Readings are easily made from left to right.

Sir Wm. Thomson and Mr. Varley's sliding resistance box, used in bridge measurement. It is so arranged as to enable the operator to vary the two arms of the bridge by causing a handle to slide over successive contacts arranged round the circumference of a circle. One of the arms contains 100 coils of 20 ohms each. The other, 101 coils of 1,000 ohms each. The sum of the resistances of the two arms is constant; their ratio only is made to vary. The third arm has a fixed known resistance; the fourth is the resistance to be measured.

Precautions to be observed in using resistance boxes.—The plugs of the box must be kept perfectly clean. It is a good plan, before commencing a series of experiments, to rub the plugs over with a fine file or emery paper, taking great care that none of the emery dust adheres to the metal. In putting in a plug, it should be slightly twisted, to ensure a good contact; but this ought to be done gently, so as not to damage the ebonite. In taking out or putting in a plug, care must be taken not to disturb the plugs in its neighbourhood. Before beginning, care should be taken that all the keys make a good contact, and that they are free from grease. The metal part of the plugs should never be touched with the fingers. They should always be taken hold of by the ebonite handles.

Wheatstone's bridge.—An arrangement devised by Christie, the principle of which is represented in the adjoining figure:

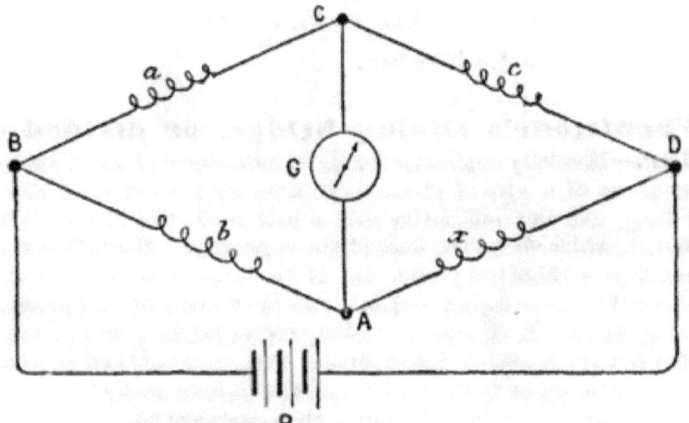

Fig. 13.—Principle of the Wheatstone's Bridge.

Four resistances $a\ b\ c\ x$ form the four sides of a lozenge A B C D. A battery is connected to the two ends of a diagonal BD. A galvanometer G is connected to the ends of the diagonal AC. When the galvanometer is not deflected, there is the following relation between the resistances $a\ b\ c\ x$:

$$\frac{a}{b} = \frac{c}{x}.$$

This relation is a consequence of Kirchoff's laws.

Wheatstone applied Christie's arrangement to the measurement of resistances, so that it is generally known under the name of Wheatstone's bridge or Wheatstone's balance.

When a and b are known, if c is varied until the galvanometer comes to zero, the value of x is:

$$x = c\frac{b}{a};$$

a and b are the *arms* of the bridge. When a and b are equal, $x = c$.

By varying the ratio between a and b resistances may be measured which are either much greater or much smaller than the resistances to be found in the resistance box. Thus the Post-Office bridge box, which has two arms of 10, 100, and 1,000 ohms, enables measurements to be made of resistances which vary from $\frac{1}{100}$ of an ohm up to 1,000,000 ohms, which is quite sufficient for most practical work. Boxes which have arms containing resistance coils of 10,000 ohms enable measurements to be made from $\frac{1}{1000}$ of an ohm up to ten megohms.

Wheatstone's sliding bridge, or divided wire bridge.

—Specially applicable for the measurement of small resistances. It is made up of a wire of platinum iridium alloy, or German silver, one mètre long, and one millimètre and a half in diameter, stretched on a long board, which forms the base of the apparatus. Beneath it is a scale graduated in millimètres; each end of the wire is soldered to a metal block joined to large copper bands. In one of the arms is placed a fixed known resistance R (a standard ohm, for example); one of the wires from the battery is attached to a sliding contact, which can be run along the wire. The other battery wire and the galvanometer are connected as in the ordinary bridge. Calling x the resistance to be measured, we have

$$x = R\frac{a}{b}.$$

It is only necessary to know the ratio $\frac{a}{b}$, and not the absolute value of the resistances a, b. If the wire is homogeneous and of uniform diameter, this ratio is that of the lengths on each side of the slider; the slider generally has a vertical rod which is kept up by a spring; the battery wire is attached to this rod, and contact is established only at the desired moment by pressing down the spring. This apparatus, which is very useful for certain kinds of work, is not very accurate, because it is based on the hypothesis that the wire, in spite of exposure to the air, has a uniform resistance. This is not strictly true, on account of oxydisation and the abrasions made by the slider.

Thickness of wire for resistance coils.—The wire ought to be larger as the coil represents a smaller resistance. The following table shows the mean thicknesses, according to Mr. Sprague:

Coils of 1 ohm,	No. 18 to 21, B.W.G.	.	1·24 to ·31 mm.
,, 10 ,,	No. 20 to 29,	,,	.	0·89 to ·33 ,,
,, 100 ,,	No. 25 to 34,	,,	.	0·5 to ·17 ,,
,, 1,000 ,,	No. 32 to 40,	,,	.	0·22 to ·1 ,,

Resistance coils are generally made of German silver, of which the specific resistance varies considerably with the proportion of nickel; as a first approximation and as a guide in purchasing wire we may take it that

1 mètre of German silver wire, 1 mm. in diameter, has a resistance of ·27 ohm.
1 ,, ,, ,, ,, ·5 ,, ,, ,, 1·08 ,,
1 ,, ,, ,, ,, ·1 ,, ,, ,, 27 ,,

at the temperature of 0° C. Different samples vary between the resistances given above and the double of them.

STANDARDS OF ELECTROMOTIVE FORCE.

Electromotive forces are measured either by the aid of instruments so graduated that they give the value directly in volts, or by comparing them to other electromotive forces taken as standards. There is no standard in existence having an e. m. f. exactly equal to one volt. In practice elements are used so arranged that they give a constant e. m. f. of known value. We will describe here those standards which are most commonly employed, stating for each one of them the conditions necessary in order that their e. m. f. may be considered as practically constant.

The Post-Office standard cell is a form of Daniell's cell. It is composed of a box (Fig. 14) containing three distinct receptacles; the left-hand one contains a plate of zinc Z immersed in water, the right-hand one a flat rectangular porous pot C, containing a plate of copper, and filled with a saturated solution of copper sulphate. The porous pot is immersed in water. These two receptacles are only used when the cell is at rest. The centre one contains a half-saturated solution of zinc sulphate; at the bottom of it is a little zinc cylinder x, in a special compartment. When the cell is to be used the porous pot is taken out from the place in which it remains when the cell is at rest, and placed in the centre compartment, in which the zinc is also placed. The cell is then ready for use; both the zinc and the porous pot are taken out and put in their resting compartments when the cell is done with. The small quantity of sulphate of copper which has passed through the porous pot whilst the cell is in action is reduced by the little zinc cylinder x. The zinc sulphate solution is thus kept pure.

Under good conditions a newly put up cell gives an e. m. f. of 1·079 volts. In ordinary working the officials of the post-office assume that it has an e. m. f. equal to 1·07 volts.

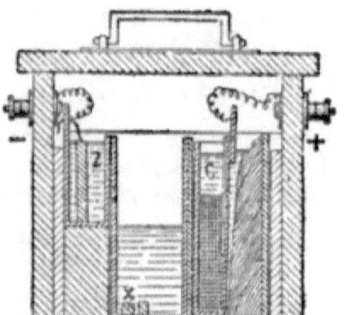

Fig. 14.—Post-Office Standard Cell.

Mr. Warren De La Rue's chloride of silver battery.—Each element is composed of a glass vessel closed by a paraffin stopper. The negative plate is chemically pure zinc, not amalgamated; the positive plate is chloride of silver enclosed in a cylinder of vegetable parchment. The chloride of silver AgCl is cast round a flattened silver wire, which serves as an electrode. The exciting solution is composed of 23 grammes of chloride of ammonium (sal-ammoniac) to a litre of water. The paraffin is melted down on to the edge of the glass, and round the wire, by means of a hot iron. The e. m. f. of this cell is 1·068 volts. The battery acts better the more often it is used. If it remains long at rest a very adherent coating of oxychloride of zinc is formed, which greatly increases the resistance of each element.

With bromide of silver the e. m. f. is . . . ·903 volt.
With iodide of silver ,, ,, . . . ·758 ,,

Mr. Latimer Clark's standard cell can only be employed for those measurements in which the battery produces no current (Law's method, for example). At a temperature of 15° C. the e. m. f. is 1·457 volts. Increase of temperature diminishes the e. m. f. by ·6 per cent. per degree centigrade for ten degrees above and below 15° C. In this cell the positive element is pure mercury, which is covered by a paste formed by boiling sulphate of mercury in a saturated solution of sulphate of zinc; the negative element is formed by zinc which has been purified by distillation; it rests on the paste. The best way of putting up this cell is to dissolve sulphate of zinc in boiling distilled water to saturation. When it has grown cold the solution is decanted from the crystals, and is mixed with pure sulphate of mercury until a thick paste is obtained, which is then boiled in order to get rid of the air. This paste is then poured on to the mercury, which has been previously heated in a suitable glass vessel. A piece of pure zinc is then suspended in the paste. It is as well to close the vessels hermetically with melted paraffin wax. Contact is made with the mercury by a platinum wire, which passes through a glass tube fixed to the interior of the vessel, and which plunges below the surface of the mercury; or, better still, through a little glass tube blown on to the side of the vessel, and opening close to the bottom of it.*

Mercurous sulphate (Hg_2SO_4) can be obtained commercially. It may also be prepared by dissolving excess of mercury in sulphuric acid raised to the point of ebullition. The salt, which is an almost insoluble white powder, must be washed with distilled water. Care must be taken to obtain it quite free from mercuric sulphate (bi-sulphate), the presence of which is shown by the mixture turning yellow when water is added. It is essential to wash the salt carefully, as the presence of free acid or bisulphate produces a change in the e. m. f. of the cell.

Zinc cadmium element.—An amalgam of zinc is placed in a vessel and covered with a solution of pure sulphate of zinc; in another vessel is placed an amalgam of cadmium covered with a solution of sulphate of cadmium; the two vessels are joined together by a capillary syphon; platinum wires connected with the amalgams form the poles of the element. The e. m. f. is ·28 volt (Debrun); the calculated e. m. f. is ·35 volt. This standard answers for all measurements made with condensers.

Simple standard cell.—A plate of zinc and a plate of copper immersed in a saturated solution of sulphate of zinc, give an e. m. f. of one volt (*Ayrton* and *Perry*); this cell must only be employed with a condenser to prevent polarisation. A very good standard for ordinary measurements is a common Daniell cell, with a very tall porous pot. The zinc and zinc sulphate solution are in the porous pot, which is well soaked in melted paraffin wax for two or three inches from its upper border. If the zinc sulphate solution be always kept some inches higher than the copper sulphate solution in the outer pot, the cell may be left at rest for a long time without any deposit of copper taking place on the zinc. Such a cell may be trusted to give an e. m. f. within 3 or 4 per

---

* Experience shows that these cells often vary very much after the lapse of a few months. This is probably due to bad insulation; they should be very carefully sealed and kept in a dry place.

cent. of 1·07 of a volt for many weeks, and even longer if it be kept on closed circuit through a resistance of a few thousand ohms when not in use.

STANDARDS OF CAPACITY.

Those employed in practice are *condensers*, of which the capacity varies between one-third of a microfarad and ten microfarads. In scientific researches absolute air condensers are sometimes used, of which the capacity is calculated according to their geometrical form, but up to the present time they have not been used for general practical work.

Condensers.—A condenser is a Leyden jar of large surface and small volume.

Varley's condensers are made of very thin silver leaves or sheets of tinfoil, covered with paraffin.

Latimer Clark's condensers are made with tinfoil and sheets of mica covered with paraffin or gum-lac.

Willoughby Smith's condensers have a special kind of guttapercha containing a large quantity of gum-lac as a dielectric.

The capacity of a condenser C is measured in microfarads. It is proportional to the specific conductive capacity K of the dielectric to the area of the opposed surfaces S, and inversely proportional to the thickness of the insulating layer d:

$$C = \frac{KS}{2\pi d}.$$

A one-microfarad condenser contains about 300 circular leaves of tinfoil, separated by leaves of mica and contained in a box 8 centimètres high and 16 centimètres in diameter. When the condenser is not in use the two armatures must always be kept connected by a plug which keeps the condenser discharged.

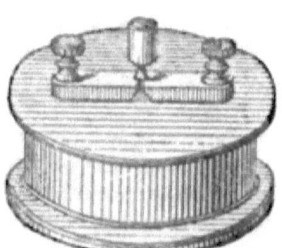

Fig. 15.—Half-Microfarad Condenser.

Some standard boxes of condensers are composed of fractions of a microfarad which may be grouped at will by the aid of plugs so as to vary the capacity and facilitate measurements.

Fig. 16 shows the upper part of a condenser of one microfarad divided into five parts as follows:

·05   ·05   ·2   ·2   ·5.

Construction of one-microfarad condenser (*Culley*).—Requires 37 sheets of good tinfoil, 184 millimètres by 152, separated from each other by two leaves of very thin hot-pressed paper, such as is used for bank notes. The two series are composed respectively of eighteen and nineteen sheets of tinfoil. The one with nineteen leaves forms the exterior of the condenser, and is connected to earth. The

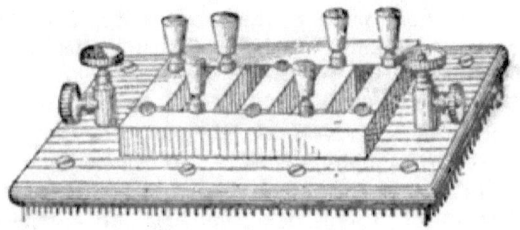

Fig. 16.—One Microfarad divided into Five Parts.

additional leaf has the effect of neutralising the effects of the induction of neighbouring objects. The paper should be thoroughly dried, and soaked in paraffin, either by immersing it in a bath of melted paraffin or painting it over with a camel-hair pencil.

In order to construct this condenser a plate of sheet iron is taken, a little larger than the sheets of paper, and mounted on four legs, so that it may be heated from below by means of a gas jet. Its surface ought to be plain, and polished, with a groove round the edge to receive the excess of paraffin. The paper is cut out in sheets large enough to stick out beyond the sheets of tinfoil about 25 millimètres all round. The two upper corners of each sheet are turned up, one of the corners of the metallic sheets is also turned down. They are spread out with care, and joined into two series, one of eighteen and the other of nineteen sheets, by soldering together the corners which are not turned up opposite to the turned up corners of the same side of the sheets, so as to make two distinct books of them. A sheet of the paper is laid on the warm plate of sheet iron; it is covered with melted paraffin with a very silky camel-hair pencil; on this is laid the first tinfoil sheet of the book containing nineteen sheets. This sheet is then covered with varnish; on it is placed two sheets of the paraffined paper. On this paper is placed the first sheet of the series of eighteen sheets, so that the soldered corners correspond to the turned up corners of the paper, and are opposite to the soldered corners of the other series. A coat of varnish and two sheets of paper are then added as before; the second sheet of the series of nineteen is then spread out, taking care to smooth each leaf carefully as it is put

F

into its place. When the apparatus is constructed it is placed between two hot metal plates and submitted to a pressure of 400 kilogrammes to squeeze out the excess of paraffin and make the whole compact.

By this means the alteration in the capacity of a condenser is avoided, which would be produced by any alteration in the distance between the metal plates. Two thicknesses of paper are used in order to ensure good insulation, which might be destroyed if there were any small hole or break in the paper. Care must be taken to arrange a battery of one to ten elements, and a galvanometer between the two series of metallic plates so as to observe whether the insulation remains perfect whilst the condenser is being built up. When the condenser has become quite cold its capacity is verified by some suitable method. If the capacity is too small it is increased by applying pressure. If this method is insufficient more tinfoil leaves must be added; if, on the other hand, the capacity is too large, a few tinfoil leaves are removed. When the operation is finished, the condenser is placed between two wooden mounts joined by two wooden screws, which keep up an invariable pressure; the whole is then placed in a box carrying two binding screws connected to the two armatures of the condenser. The same method is employed for mica condensers, but the elasticity of this substance causes their capacities to vary with the pressure to which they are subjected. For equal volume a mica condenser has a larger capacity than a condenser with paraffined paper, because the inductive capacity or inductive power of mica is greater than that of paper.

ACCESSORY APPARATUS FOR ELECTRICAL MEASUREMENTS.

Circuit breakers and commutators are used to make and break electrical connection between the different measuring

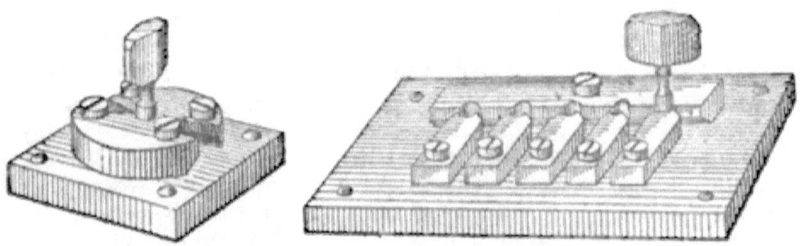

Fig. 17.—Plug Circuit Breaker.   Fig. 18.—Battery Commutator.

apparatus. They are made both with keys and with plugs, and for one or more directions as may be required. The form known under the name of *battery commutator* is a very convenient form when it is necessary to vary the number of elements in circuit rapidly.

Reversing commutators.—The simplest form is composed of four thick brass quadrants screwed to a plate of ebonite, and not touching each other, having semicircular slots cut out in them, in which brass plugs may be introduced to make electrical contact between the quadrants.

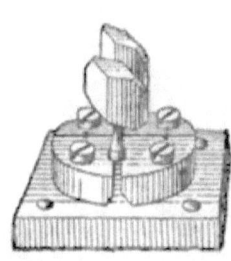
Fig. 19.—Plug Reversing Commutator.

Fig. 20.—Reversing Key.

Reversing keys are used to connect the galvanometer to other measuring apparatus. The contacts are so arranged that when one of the springs is pressed down, the current flows in one direction, and when the other is pressed down, in the reverse direction. Two other keys enable the springs to be wedged, so that permanent contact can be maintained when necessary.

Short-circuit key or tapper is connected between the terminals of the galvanometer, to prevent currents of too great strength accidentally passing through the coils. In its normal condition the spring presses against a platinum contact; and when pressed down, against an ebonite contact. A movable stop enables the key to be kept down permanently when it is necessary.

Fig. 21.—Short-Circuit Key or Tapper.

Discharging key, used in measurements with condensers. The most commonly used is *Sabine's*. It has three keys. The first puts the condenser in circuit; the second insulates the condenser; the third connects the condenser and the galvanometer.

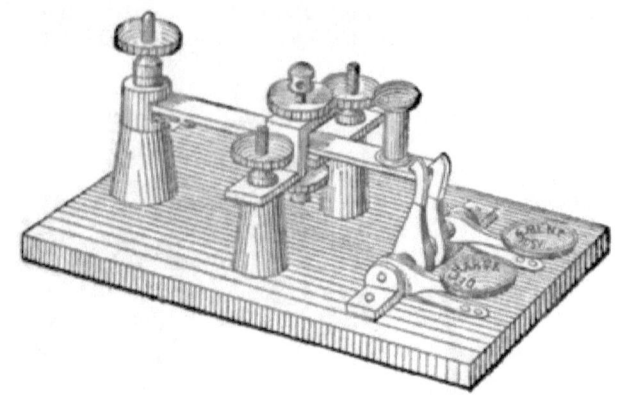

Fig. 22.—Sabine's Discharging Key.

Double contact key is used in bridge measurements with ordinary bridge boxes. The connections are so arranged that the first contact closes the battery circuit, and the second the galvanometer circuit almost immediately one after the other, and this is effected by pressing down one key only.

GENERAL METHODS OF MEASUREMENT.

Methods of measurement may be divided into two great classes:

(1) *Direct methods*, in which a quantity to be measured is compared to a quantity of the same kind by one of the three following methods:

    a. By opposition.
    b. By substitution.
    c. By comparison.

(2) *Indirect methods*, in which the magnitude of the quantity to be measured is deduced from the value of two or more other known quantities by means of a known relation. (Example: Heat disengaged in a voltaic arc, resistance of a conductor, when the strength of a current passing through it and the difference of potential between its two extremities are known.)

a. **Differential, zero, equilibrium or balance, opposition methods.**—These consist in opposing the unknown magnitude by a known magnitude and reducing to zero, or compensating the effect of the unknown magnitude by variations of the known magnitude. When equilibrium is established, the equality of the magnitudes is shown by the equality of the effects. In this case we have to observe the non-existence of a phenomenon. The instrument does not require a scale, but a variable standard or graduated standards are necessary. The type of this method is the ordinary process of weighing. In electrical methods, we may take the Wheatstone bridge method as a type. In this it is only necessary to be satisfied of the equality of potential between two given points; for this purpose galvanoscopes, galvanometers, electrometers, etc., may be employed. The accuracy of the measurement depends upon the sensibility of the instrument which shows the equality of the potential between the two points under consideration.

b. **Substitution methods.**—The effect produced by the quantity to be measured is noted, and a known magnitude, capable of producing the same effect, is substituted for it. The instrument of observation must be graduated. The graduation may be arbitrary, but again it is necessary to have a variable standard or graduated standards. Sometimes one or both of the two effects is reduced in a known proportion, so as to bring the indications within the limits of the scale, or to that part of the scale at which the instrument is most sensitive.

c. **Comparison methods.**—First of all, the effect is measured of a known fixed magnitude, then that of an unknown magnitude. The ratio of these magnitudes is deduced from the ratio of their effects.

A calibrated measuring instrument is necessary, or a fixed standard, but these are not necessary if we know the constant of the instrument, and the relation which connects the magnitude to be measured with the readings of the instruments. (Example: Sine galvanometer, tangent galvanometer, etc.)

MEASUREMENT OF CURRENTS.

Currents are measured by their electro-magnetic, electro-dynamic, or electro-chemical actions. There are these three classes of measuring instruments:

(1) *Galvanometers*, based on electro-magnetic action.

(2) *Electro-dynamometers*, based on the action of currents or electro-dynamic action.

(3) *Voltmeters*, based on chemical action.

(1) GALVANOMETERS.

Every apparatus in which a magnetised needle is deflected by a current forms a *galvanometer*. A galvanometer is a *galvanoscope* when it indicates the passage of a current without measuring it. In zero methods galvanometers act as galvanoscopes. Galvanometers are based on a discovery made by Œrsted, in 1819, and on Schweigger's multiplier. The application of astatic needles to galvanometers is due to Nobili, and that of the reflecting mirror to Sir Wm. Thomson. Galvanometers vary almost infinitely both in form and arrangement. We will only describe those most in use, but we will first point out the general principles which have been applied to them, and which, when they have a considerable importance, give the name of the instrument.

Absolute galvanometer.—Enables the strength of a current to be directly measured in terms of the dimensions of the galvanometer and of the horizontal component of terrestrial magnetism.

Astatic galvanometers.—The directive force of the earth is diminished, so as to increase the sensibility of the instrument, either by means of a directing magnet or by a pair of needles forming an astatic system.

Balistic galvanometer.—The measurement is made by the impulse on the needle produced by the action of a momentary current.

Calibrated galvanometer.—The graduation of the scale is made, not in degrees, but in terms of the strength of the current, or the ordinary gradation in degrees is accompanied by a reduction table.

Current galvanometers.—A galvanometer in a circuit which measures directly the strength of the current passing.

Dead-beat galvanometer.—The needle goes to its position of equilibrium almost without vibrating.

Differential galvanometers measure the difference of the action of two currents on a magnetised needle.

E. m. f. galvanometer.—A galvanometer formed of relatively fine wire, placed as a shunt between two points of a circuit, the difference of potential between which is to be measured (indirect measurement).

Fine wire or *tension galvanometer.*—These are unscientific names expressing the nature of the wire with which the coils are wound. This nomenclature is disappearing and giving place to the indication of the resistance of the galvanometer in ohms.

Mirror galvanometer, or *reflecting galvanometers.*—The index is formed by a ray of light.

Quantity galvanometers have thick wire.

Sine galvanometer.—The law of deflection is connected with the sine of the angle of deflection.

Tangent galvanometers.—The law of deflection is connected with the tangent of the angle of deflection.

Torsion galvanometer.—The action of the current is balanced and measured by the torsion of a rod or wire.

It may be seen by this enumeration that the name of a galvanometer only defines its principal property, and that any given galvanometer may present several different properties to the same degree. The choice of an instrument to be used depends upon the nature of the measurements to be performed, the desired degree of accuracy, and the kind of people who have to make use of it, etc. Galvanometers are also sometimes called multipliers or rheometers, but these terms are disappearing.

Sine galvanometer is composed of a vertical galvanometer coil, in the centre of which is placed a magnetised needle, free to turn in a horizontal plane. The plane of the coil is placed in the plane of the magnetic meridian, *i.e.* parallel to the needle. The current is then passed through the coil; the needle is deflected; the coil is then turned in the direction of the deflection until the needle again becomes parallel to the coil. The strength of the current is proportional to the sine of the angle through which the galvanometer coil has been turned, whatever may be the relative dimensions of the needle and the coil, and whatever may be the shape of the coil. Sine galvanometers are very sensitive, because the coil may be very close to the needle, but the observations require more time than those made with other instruments; each movement of the coil producing a deflection of the needle, some time is required in order to get the needle and coil parallel. The sensibility increases with the deflection.

Tangent galvanometer.—When a very small magnetised needle is placed in the centre of a circular galvanometer coil of very large dimensions, and the axis of the needle is in the plane of the coil, it may be shown that the tangent of the angle of deflection is sensibly proportional to the strength of the current passing. The maximum sensibility is at $0°$; the sensibility disappears at a deflection of $90°$. It is necessary, therefore, to measure very small angles, and to compensate for the smallness of the deflection by the accuracy of its measurement, or by the employment of Thomson's reflecting galvanometer (page 73).

Gaugain's conical multiplier.—A galvanometer coil

forming a short frustum of a right cone, at the apex of which the needle is placed; the height of the cone is equal to half the radius of the base. Helmholtz has shown that the effect is doubled by having two coils placed symmetrically on each side of the needle; this forms the tangent galvanometer which approaches most nearly to the theoretical conditions. For a galvanometer with one coil we have:

$$C = \frac{a}{4 \cdot 504 n} H \tan \delta.$$

C represents the strength of the current; δ the angle of deflection; a the radius of the base; n the number of turns in the coil; H the horizontal component of the earth's magnetism. In order to get an absolutely uniform magnetic field at the centre of the needle, three vertical parallel coils must be used, the centre one larger than the other two, so that all three lie on the surface of a sphere of which the small needle occupies the centre.

Post-office tangent galvanometer, for telegraphic *measurements*, is formed of a circular ring of brass 15 centimètres in diameter, on which the coils are wound; the needle is about 18 millimètres long, which practically gives sufficient accuracy. The needle carries, at right angles to its axis, an index 12 centimètres long, which moves above a scale with two graduations, one in degrees on one side of the ring, the other in tangents on the other side. To prevent errors of parallax in reading, a piece of looking-glass is placed in the plane of the graduated ring, which reflects the index; when a reading is taken, the image of the index must be concealed by the index itself. The ring carries three coils, one composed of only three turns of thick wire, the other two have each a resistance of 25 ohms. At pleasure any one of the coils may be used separately, or two coils in series, or two coils parallel, according to the nature of the currents to be measured.

Schwendler's tangent galvanometer, used in the Indian telegraph service. This is a tangent galvanometer with two coils, one of 1 ohm and the other of 100 ohms. The one-ohm coil is accompanied by two resistances of 20 and 200 ohms, which may be added to the circuit. The one-hundred ohm coil has also two similar resistance coils of 1,000 and 2,000 ohms. The instrument has also a reversing key, so that readings may be taken on both sides of the scale; two plugs to introduce one or the other of the coils into the circuit, and two terminals for the attachment of the wires. For the measurement of very powerful currents the copper ring which supports the coils is cut in two, and its extremities connected to two other terminals, and thus forms a third coil.

The length of the needle is not more than one-fifth of the diameter of the coil. This needle carries an aluminium index fixed at right angles to its axis, and provided with wings of the same metal to check its oscillation. The whole is enclosed in a cubical box of 15 centimètres side. The closing of the box automatically raises the needle, and disengages it from the pivot. With one Daniell's cell the 100 ohms coil and 2,000 ohms resistance in the circuit, the deflection is $5°$; one half of the scale is graduated in degrees, the other half in tangents.

Obach's galvanometer.—A tangent galvanometer, the coils of which can be inclined from the vertical position. When the coils are inclined at an angle θ, if δ be the deflection, the current is proportional to

$$\tan \delta \sec \theta.$$

The angle through which the coils are turned is read on a divided quadrant by means of a vernier. The instrument is provided with one coil of very low resistance, and one of very high resistance, so that it can be used both as an "ammeter," and a "voltmeter"; the power of varying the constant of the instrument by inclining the coils gives it a very wide range of utility, enabling both small and large currents and differences of potential to be measured by one and the same instrument.

Siemens' universal galvanometer.—An instrument in which a set of resistance coils, a wire Wheatstone-bridge, and a galvanometer with movable coils, which forms a sine galvanometer, are included in one piece of apparatus. All telegraphic measurements of resistance of e. m. f. and current strength can thus be performed with the same piece of apparatus, the wire bridge being convertible into a Clark's potentiometer.

Sir William Thomson's reflecting galvanometer.—This is the most sensitive apparatus known for measuring small currents and high resistances. It varies very much in form, but in principle it is composed of a light magnetic needle suspended in the centre of a large coil of wire; and of a reflecting system which enables the deflections of the needle to be amplified. A long index is formed by a ray of light reflected upon a divided scale by a little mirror, which is attached to the magnetic needle. The deflections always being very small, and the coil relatively large, the deflections are always sensibly proportional to the strengths of the current. It is constructed either astatic, dead-beat, or differential, etc.

In the non-astatic form it is composed of four little magnets from 4 to 5 millimètres in length, cemented to a small mirror. The diameter of the mirror is about 6 millimètres, and the total weight of the mirror and needle together is not more than 7 centigrammes. The object of multiplying the number of needles is to obtain the maximum of magnetisation with the minimum of mass, because the needle comes more readily to zero as its magnetisation is greater.

The mirror is suspended by a cocoon fibre, and placed in the centre of a coil enclosed in a cylinder of brass. The front face is closed by a plate of glass. The cylinder is supported on a tripod with levelling screws, by which the instrument can be levelled. A slightly curved directing magnet, supported by a vertical rod fixed on the case of the instrument, forms an artificial meridian, of which the strength and direction can be varied by causing it to slide up or down, or turn round on the rod, so as to act with more or less force upon the suspended magnet, and so vary the sensibility of the instrument.

Regulation of the sensibility of the galvanometer.—When the poles of the directing magnet are arranged like those of the terrestrial magnet (the marked pole to the south) its directive force is added to that of the earth, and the sensibility of the apparatus diminishes. By turning this magnet through 180°, its directive force is opposed to that of the earth, and the sensibility increases. To get the maximum of sensibility the magnet is lowered until the two actions neutralise each other, then it is slightly raised so as to preserve a slight directive force to bring the luminous index back to zero.

Thomson's astatic galvanometer.—The astatic system is only employed in instruments wound with a very long wire, each needle separately being surrounded by a coil, and the current passing in opposite directions in each coil. As the whole system is rather heavy the lower needle has attached to it a small lozenge-shaped plate of aluminium to damp its vibrations; the adjustment of the position of the directing magnet is controlled by means of a tangent screw. The square or cylindrical box enclosing the apparatus has an opening to contain a thermometer and is also provided with a spirit level, which enables the system to be put in a perfectly vertical position. Each of the coils is composed of two parts separated by a vertical plane. The astatic system can thus be withdrawn for adjustment or repair of the cocoon fibre, etc. There are thus really four distinct coils, which are in connection with eight terminals placed on the foot plate. According to the way in which these terminals are connected the coils may be grouped in series, parallel, or with two coils parallel, and the two systems of two coils in series, so that the resistance of the galvanometer can be varied according to the measurements which have to be made. The shunts which accompany a galvanometer always correspond to the arrangement of the coils in series. The cocoon fibre is attached to a knob which can be raised or lowered at pleasure. When it is lowered the needles rest on the coils, and the instrument may then be removed without risk of breaking the fibre. The readings of the deflections of the galvanometer are made by means of a lamp, divided scale, and movable mirror.

The arrangements being the same for all reflecting apparatus we will describe it once for all.

Lamp, scale, and mirror.—An arrangement devised by Sir William Thomson for observing and measuring very small angular deflections, which is applied to all reflecting apparatus, galvanometers, electrometers, etc., and in which a ray of light acts as a long index without weight, and therefore without inertia. A movable mirror is fixed to the apparatus of which the deflections are to be measured, a lamp and scale are placed before it at a variable distance, generally 60 to 80 centimètres, the light of the lamp passes through a narrow slit cut in the base of the scale, falls on the movable mirror, is reflected from it, and returns, forming a small luminous image on the upper part of the scale; the least movement of the mirror displaces the image along the scale. The distance passed over is equal to that which an index of double the length of the distance from the mirror to the scale would pass over.*

The opening is sometimes a slit, the movable image is then a luminous vertical line; sometimes a circular hole crossed by a very fine platinum wire, stretched vertically, when the image is an illuminated circle crossed by a thin black vertical line. If the mirror is plain the light is converged so as to come to a focus on the scale by means of a lens. Sometimes the mirror is concave, which does away with the necessity for a lens. The concave mirrors being very expensive, a thin disc of silvered glass, and a lens are more often used; microscopic covering glass answers very well for this purpose. The scale is generally divided into millimètres, and printed in black on white glazed paper; sometimes it is formed of ground glass, and the deflections are read on the other side of the scale. When a petroleum lamp with a flat wick is used, the wick ought to be placed with its edge turned towards the slit. An incandescent electric lamp gives a very sharp image with a slit.

Hole, slot, and plane.—An arrangement devised by Sir

---

* In accurate scientific measurements this distance is sometimes 6 mètres, so the deflections read upon the scale are equal to those of an index 12 inches long. The deflections being infinitely small, the strengths are exactly proportional to the divisions of the scale.

W. Thomson, by which any piece of apparatus, galvanometer, electronometer, etc., which rests on a table by means of three levelling screws, can be removed and always replaced in the same position. The three legs are numbered 1, 2, and 3. Foot No. 1 is placed in a small hole made in the table; foot No. 2 in a short slot, whose axis, when prolonged, passes through the hole; foot No. 3 on the plane of the table. All error is thus avoided when the instrument is replaced after being moved. The advantage of having a slot instead of a second hole is that the arrangement allows of more than one instrument being used in the same place, independently of the dimensions of the instruments, if the feet of the levelling screws be all of the same diameter. By this method, therefore, one lamp and scale will answer for several instruments.

Sir William Thomson's ship's galvanometer. —For cable testing on board ship. The mirror and needle are attached to a stretched wire fastened at both ends, and passing through the centre of gravity of the system to prevent oscillations caused by the rolling and pitching of the ship. The apparatus is further surrounded by a thick iron cage in order to preserve the instrument from disturbances produced by external magnetic forces; a strong directing magnet of horse-shoe shape embraces the coils and directs the needles. The spot of light is brought exactly to zero by means of a small regulating magnet worked by a rack, pinion, and milled head placed behind the galvanometer.

Dead-beat galvanometer.—A galvanometer is called dead-beat when it rapidly takes up its position of equilibrium under the action of a current, and comes rapidly to zero when the current is interrupted. This result is obtained by many details of construction. These are the most used:

(1) Surrounding the needle with a mass of copper, which damps the vibrations by the effect of the induced currents which the movement of the needle produces in its mass (*Weber*).

(2) The needle is provided with a light vane, which moves in water or in air, and resists sudden movements.

(3) The needle has a very small mass, and strong magnetisation, and very large directive force is applied.

Sir William Thomson's dead-beat galvanometer.—The numerous oscillations of the mirror in ordinary galvanometers often cause precious time to be lost in making measurements. Sir W. Thomson's dead-beat galvanometer gets over this difficulty. It is a modification of the ordinary non-astatic reflecting galvanometer. The centre of the coil is occupied by a brass tube A, of such length that the part *ab* is in the middle of its length, the tube *a* is closed by a small plate of glass, it is tapped at one end, on which a small ring *c* is screwed, into which a third part of the tube, also closed by a plate of glass, is screwed so as to form a completely closed air-chamber. A small mirror *m*, carrying a small magnetised needle, is placed in the centre of the tube *c*; the mirror is very nearly of the same diameter as the tube, only just having clearance. It is suspended by an extremely short cocoon fibre; the space *ab*, closed by a small glass plate, is just deep enough to enable the mirror to give a good deflection on the scale. By this arrangement all violent movement by the action of the current is prevented;

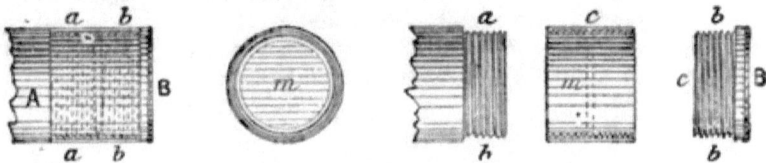

Fig. 23.—Sir W. Thomson's Dead-Beat Reflecting Galvanometer.

instead of passing the point of rest, and coming back again, the spot of light travels slowly to its proper position, and stops there without passing it. When the current is interrupted the spot of light comes back to zero. The suspension being a very short fibre, the mirror does not move so freely as in the ordinary galvanometer, its sensibility is therefore not so great, but it is nevertheless quite sufficient for most purposes. It is easy to replace the fibre when it is broken. One end of the fibre being fastened to the mirror, the other end is passed through a little hole bored in *c*. The fibre is then stretched until the mirror is suspended, and does not touch the sides of the tube. A drop of varnish is then let fall upon the hole, which is thus closed, and the fibre fixed (*Kempe*).

Marcel Deprez' dead-beat galvanometer.—A very light soft-iron needle placed between the two poles of a strong horse-shoe magnet. The index is made of straw, hair, or aluminium. The deflections are produced by two coils of coarse or fine wire, according to the currents to be measured, placed on each side of the needle. The mathematical expression of the law, which connects the current strengths with the deflections, is not known. In some arrangements the deflections of the needle are amplified by a cord and pulley arrangement.

Ayrton and Perry's dead-beat galvanometer.—Something like the preceding apparatus, but the shape of the coils and the form of the pole pieces of the magnet are so calculated that the deflections are proportional to the current strengths up to an angle of about 40°.

These instruments are wound both with coarse and fine wire, so as to form ammeters or voltmeters. The wire is in some cases made up into a cable, the strands of which can be arranged either parallel or in series, by means of a special commutator; when the wires are parallel the constant of the instrument is one-tenth of the constant when the wires are in series. Those instruments which have this arrangement are also provided with a resistance coil, which can be thrown into the circuit by removing a plug.

The ammeter is thus calibrated; the instrument is arranged with the wires in series, and put in circuit with a standard Daniell's cell; a deflection a is thus obtained; the resistance coil (in this case 1 ohm) is then unplugged, and a second deflection b is obtained; then

$$1° = \frac{ab}{E(a-b)},$$

where E is the e. m. f. of the Daniell's cell (1·079 volts), when the wires are again put parallel.

$$1° = \frac{ab}{E(a-b)} \times \frac{1}{10}.$$

The forms without commutators are calibrated by comparison with other instruments, Siemens' dynamometer being generally used.

Dead-beat galvanometer of Messrs. Deprez and D'Arsonval.—Intended for the measurement of very small currents. A galvanometer coil is suspended between the branches of a vertical horse-shoe electro-magnet by two platinum wires, which bring the current to it, and form an elastic torsion couple. A tube of iron placed in the interior of the coil between the branches of the magnet concentrates the magnetic field. The readings are made by means of Sir W. Thomson's system of lamp, scale, and mirror. When the terminals are connected by a short circuit the apparatus comes to zero without oscillation. This property makes the apparatus very useful in zero methods. It indicates a current of one-tenth micro-ampère very clearly.

Siemens and Halske's torsion galvanometer. —Intended for commercial use as a voltmeter; composed of a bell-magnet in the shape of a thimble split longitudinally along two generators diametrically opposite to each other. The poles of the magnet are formed by the two arms thus made. The magnet is fixed on a vertical axis, and turns between two coils of fine wire, through which the current passes. The action of the current is balanced by a bifilar suspension or a spiral spring placed at the upper part. One Daniell's cell produces a torsion of $15°$; a resistance coil enables the instrument to measure up to 100 volts. A graduated table gives the number of volts corresponding to each angle of torsion.

Ampère-meter, ampèrometer, or ammeter.—The name given to commercial graduated instruments, which enable the value in ampères of a current passing through them to be known by direct reading.

Voltmeter.—A galvanometer with a long fine coil, which gives by direct reading the value in volts of the differences of potential between two points of a circuit between which it is inserted as a shunt.

In reality a voltmeter also measures the strength of the current which passes through it, but as its resistance is very great as compared with the other parts of the circuit, we may consider that its introduction as a shunt between two given points of a system does not change the conditions. The differences of potential are thus proportional to the strength of the current passing through the instrument. It is necessary to give very high resistances to voltmeters, so as to prevent the heating of the wire, which would have the effect of causing the instrument to give too small deflections. It is well, in order to avoid this heating, not to allow the current to pass continuously. A small key is generally placed on the apparatus by which the circuit can be closed at the moment of taking a reading.

Precautions to be taken in the use of voltmeters and ammeters containing permanent magnets.—These apparatus must be frequently calibrated, because of the variation in the power of the magnets. It is well to put the armatures on their magnets when they are not in use, but the armature must be removed before taking a reading. A simple plan of ensuring their removal is to fasten a plate to the armature so as to hide the scale when the armature is on the magnet; any mistake thus becomes impossible. It has lately been found that the constant taking off and putting on of the armature is destructive of the magnetism of permanent magnets. Instruments which are in constant use should not have their armatures replaced. Only when an instrument is to be laid by for some weeks should the armature be put on.

Ayrton and Perry's spring ammeter.—A soft iron needle, placed almost at right angles with the axis of the coil and attached to a spiral spring. The action of the current is balanced by that of the spring. Deflections proportional to the currents can be obtained up to an angle of 45°. This apparatus acts with alternating currents. With a fine wire coil and a special graduation, the spring ammeter becomes a voltmeter. These instruments are provided with a toothed wheel and pinion arrangement, by which the deflections of the needle are amplified. Both wheel and pinion are fitted with a spiral spring. This not only tends to prevent "back-lash" when both springs are in action, but also enables the constant of the instrument to be varied to a known extent by throwing one of the springs out of action. This mechanism has been applied to the ohmmeter and arc horse-power meter of the same inventors. These instruments are not yet in practical use.

Sir Wm. Thomson's absolute galvanometer (1882).—For commercial use. Consists of a magnetometer, which is movable along a horizontal graduated scale, and a directing magnet, the magnetic moment of which in C. G. S. units is known. The sensibility is varied by using or removing the directing magnet, and by placing the magnetometer closer to, or removing it farther from, the vertical coil through which the current passes. A potential galvanometer enables measurements to be made from $\frac{1}{10}$ of a volt up to 1,000 volts without the use of auxiliary coils. The current galvanometer from $\frac{1}{100}$ of an ampère up to 100 ampères without using shunts, the correctness of which is always doubtful.

The resistance of the first instrument is more than 6,000 ohms; that of the other almost nothing. They may be set up on any circuit that has to be measured without disturbing it, and the measured quantities determined in volts and ampères by a simple arithmetical operation. The magnetic moment of the directing magnet should be frequently verified, and great care be exercised to keep it from shocks, jars, or vibrations, and far from the magnetic fields of dynamos, etc.

Let H be the horizontal intensity of the magnetic field in C.G.S. units (either with or without the magnet), d the number of divisions on the magnetometer scale, n the number of divisions on the platform scale, E the difference of potential at the terminals of the instrument. The graduations are so arranged that

$$E = H \frac{d}{n} \text{ volts.}$$

In taking a series of readings the magnetometer is fixed, and $\frac{H}{n}$ calculated once for all. Then, by multiplying d by this ratio E is obtained.

Balistic galvanometer.—When a certain quantity of electricity is instantaneously discharged through a galvanometer, if the resistance of the air to the movement of the needle be neglected, the quantity of electricity passing is proportional to the sine of half the angle of oscillation. The resistance of the air is reduced as much as possible in the balistic galvanometer. Ayrton and Perry have given it the following form: A high resistance Thomson's reflecting galvanometer has its needles removed and replaced by the following arrangement:

Forty little magnets of different lengths are prepared, and after they have been magnetised to saturation, two little spheres are constructed with them, in each one of which all the magnets are arranged in the same direction. The spheres are built up of segments cut out of a little hollow ball of lead. Both spheres are joined together by a rigid rod, so as to form an astatic combination, which is suspended in the usual way. With this arrangement great sensibility is obtained, and the air only offers a very small resistance to the movement of the needles. It has been shown that the ratio of the first oscillation to the second is only one to 1·1695, which is sufficiently close to unity to enable us to take account of the damping effect produced by the air by a very simple correction.

The extreme limit of an oscillation is called its elongation.

Approximate correction for the resistance of the air.—Let a' be the first elongation, a'' the second elongation, on the same side of zero, the approximate arc a which would have been obtained without the resistance of the air is:

$$a = a' + \frac{a' - a''}{4}.$$

Captain Cardew's ammeter.—This instrument consists in principle of two coils wound in opposite directions; one, of many turns of fine wire, the other, of one or two turns of thick wire, acting in opposite directions on a magnetised needle, which is brought to zero by means of a directing bar magnet placed on the top of the coils. The fine wire coil is of some thousands of ohms resistance; the coarse wire coil of about ·02 to ·03 ohm. The current to be measured is passed through the thick wire. The fine wire is put in circuit with a resistance box and from one to three standard Daniell's cells. The resistance box is then unplugged, until the needle is brought to zero. Then, if

G

C = current strength to be measured,
r = resistance of fine wire coil,
R = resistance unplugged in resistance box,
n = number of Daniell's cells,
K = constant of the instrument,

$$C = K \frac{n \times 1{\cdot}079}{r + R}.$$

To find K, the two coils, with a resistance box in circuit with each, are arranged in parallel arc in circuit with a dynamo or battery of very low internal resistance. A small resistance is unplugged in the thick wire circuit, and the resistance box in the fine wire circuit is unplugged, until the needle is brought to zero. Then:

If r = resistance of fine wire coil,
R = resistance unplugged in fine wire coil circuit,
r' = resistance of thick wire coil,
R' = resistance unplugged in thick wire coil circuit,

$$K = \frac{R + r}{R' + r'}.$$

In addition to the two or three turns of thick wire, these instruments are provided with a rectangle of thick copper bars, which acts as a coil for measuring very large currents. The fine wire coil and the needle move together in a groove, so as to be nearer or farther from this rectangle, the distance being observed on a divided scale. The value of K is determined for each division of the scale. This sliding action is but little used. This instrument has the advantage of not changing its constant, which also can be readily determined at any time. If the Daniell's cells be carefully put up, its indications are very trustworthy, and may be taken as true within one or two per cent. Its disadvantages are the length of time necessary to take a reading, and its want of portability, owing to the use of Daniell's cells. With care, however, it may be used as a convenient instrument for the calibration of others.

Crompton and Kapp's current and potential indicators.—To avoid the trouble of constant re-calibration, necessary where permanent steel magnets or springs are employed as the balancing force in electrical measuring instruments, Messrs. Crompton and Kapp have devised their potential and current indicators, in which electro-magnets saturated by the current which is to be measured, replace the permanent steel magnets. Both these instruments consist essentially of a coil of wire traversed by the current, and capable of

CURRENT AND POTENTIAL INDICATORS. 83

deflecting a magnetic needle against the force of an electro-magnet. In order that an electro-magnet may suitably replace a permanent one, it is necessary that its iron core should be saturated with all the varying strengths of current for which the instrument is to be used, and also that the magnetic effect due to its coils alone should be neutralised.

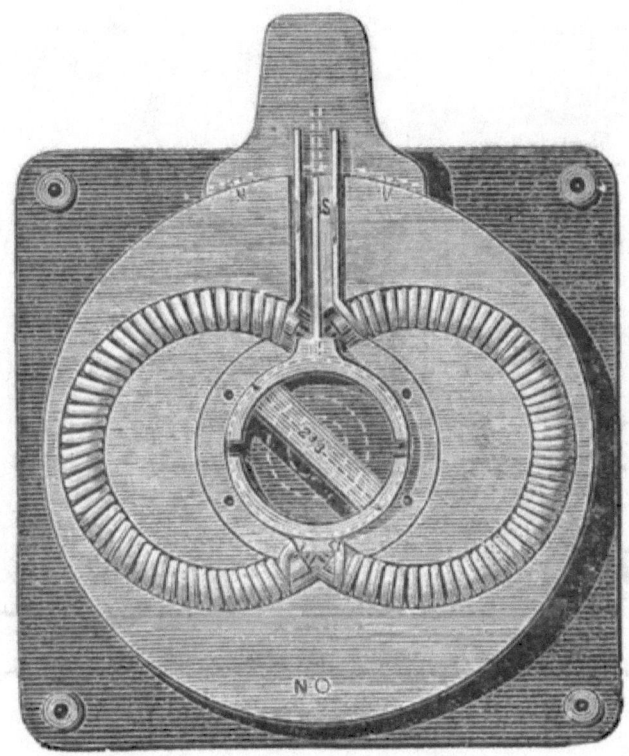

Fig. 24.—Plan of Field Magnets and Strip of Copper carrying the Current in the Current Indicator.

The first of these conditions is fulfilled by making the amount of iron in the magnets very small in comparison with that of the copper wire, and the second by setting the deflecting coil at such an angle with the line joining the poles of the electro-magnet, that while one component of the force due to it is employed to deflect the needle, the other more than neutralises the magnetic effect of the coils. As a result of this, the strength of the field actually falls off when high currents are being measured, thus allowing the increment of the angle of deflection to be comparatively large, even for high currents.

The potential indicator has a pivoted needle, swinging within a brass tube, which thus acts as a damper, rendering the instrument almost dead beat, and mounted at the lower end of a steel axle, to the upper end of

Fig. 25.—Plan of Potential Indicator, showing the two slightly inclined Deflecting Coils.

which is fastened a light aluminium pointer. The electro-magnet is of horse-shoe form, fastened to a central tubular stand, which also serves to support the two deflecting coils, one on each side; the tube within which the needle swings being inserted into the stand. The electro-magnets and deflecting coils are wound with from 50 to 100 ohms of high-resistance copper wire, and an additional resistance of German silver, nine times as great, is added. This can, however, be short-circuited by depressing a key when the instrument has to be used for measuring low electromotive forces; in this case, the value indicated by the pointer must be divided by 10. For very low readings it is preferable to read with the key depressed, as, otherwise, the very low currents produced would be insufficient to saturate the iron. A commutator allows the current to enter in the right direction, so as to bring the pointer over the scale, the handle of the commutator then points to the positive terminal.

The current indicators may have either pivoted or suspended needles. For measuring currents of 10 ampères and upwards, the deflecting coil is replaced by a single copper strip. The current entering by one of the flat electrodes splits into two parts, each part passing round the cores (wound with low-resistance wire) of an electro-magnet of horse-shoe form, the similar poles of which point towards each other. The current then unites again, and, after passing through the metal slip close under the needle, leaves the instrument by the second electrode, which is separated from the first by a narrow sheet of insulating fibre. The upper electrode is marked so as to allow the direction of the current to be easily determined.

Both instruments should be placed in such a position that the north pole of the needle points to the north, though the error caused by neglecting this is inconsiderable. The deflections in both instruments are very nearly proportional to the currents, and as re-calibration is never required, the scale of the potential indicator is divided directly into volts, and that of the current indicator into ampères. For alternating currents, the magnetised steel needle is replaced by a needle made of soft iron.

Ayrton and Perry's spring proportional am- and voltmeters.—These instruments also depend on the saturation of a small piece of soft iron by a smaller current than that likely to be measured. The directive force is obtained by a spring of peculiar form, formed of a flat ribbon of very thin metal, looking not at all unlike a very regular shaving cut in a lathe. This form of spring, when proper dimensions are given to it as regards thickness of material, length of strap, and diameter of cylinder round which it is wound, rotates through a large angle for a very small axial extension without permanent set, and the angle of rotation is directly proportional to the force tending to extend the spring. In the simplest form both am- and voltmeter consist of a hollow light iron tube closed at the bottom; this tube is suspended by the spring, the lower end of which is attached to the bottom of the tube. The tube is guided top and bottom, and to its upper edge is fastened a pointer; the whole is inserted in a coil of wire forming a sucking solenoid, the pull of the solenoid on the iron tube being proportional to the strength of the current passing (as soon as the iron tube is saturated), provided the iron core has the position determined by the inventors, and the angle of rotation of the spring, and therefore of the pointer, being proportional to the pull on the tube, the scale over which the pointer moves may (after the first few degrees) be made to show ampères and volts directly, and the same deflection will indicate one ampère or one

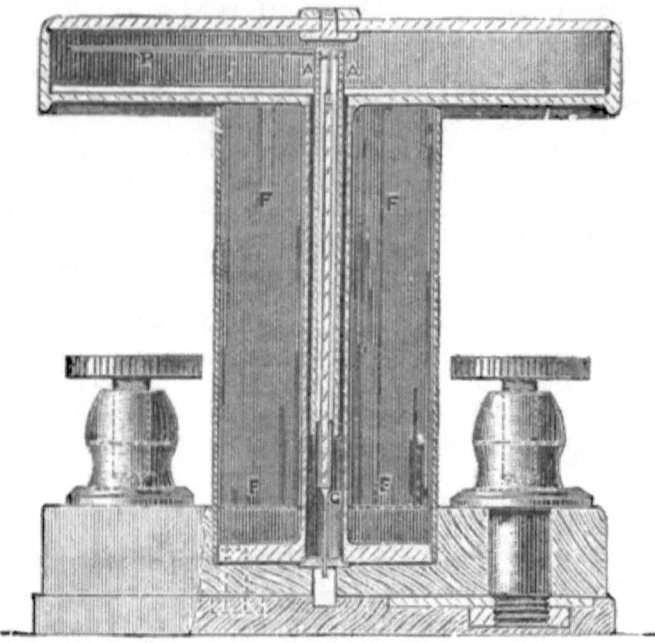

Fig. 26.—Unshielded Form of Ammeter and Voltmeter.

AA, thin soft iron tube carrying pointer P; G, spring attached to bottom of tube, and to glass cover; FF, solenoid. The figure shows the guiding pins at top and bottom of the tube.

volt at all parts of the scale. In order to avoid the labour of graduating each instrument separately, a regulating coil is provided outside the solenoid, by which the instrument can be adjusted. As soon as the adjustment is made by the maker, the coil is immovably cemented in its place. The ammeter is wound with copper strip of the same width as the reel on which it is wound, the separate layers being divided from each other by a layer of varnished fabric.

The scale of these instruments is very open, the readings being accurately proportional, between about 7° and 270°. The instruments are found to be but little affected by the magnetic field of dynamo machines,

SHUNTS AND CIRCUIT RESISTANCE COILS.

etc., and may, indeed, be used nearer to such disturbing fields than the permanent magnet instruments of the same inventors. In cases in which it is required to have instruments perfectly shielded from surrounding magnetic influences, Messrs. Ayrton and Perry have constructed ammeters and voltmeters depending on the action of the same kind of spring as that described above, but in these instruments the solenoid is replaced by a peculiar form of tubular magnet. So perfectly shielded are these

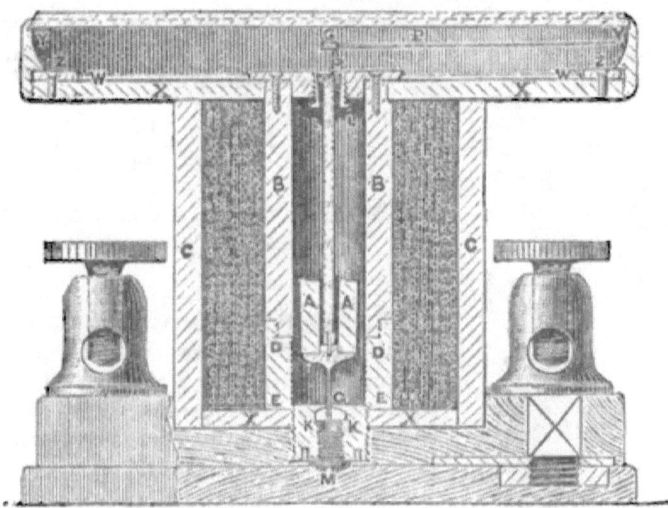

Fig. 27.—Shielded form of Ammeter and Voltmeter.

AA, Iron tube suspended by spring; GG, guiding pins; P, pointer; BB, soft iron tube, parted at D by brass or other non-magnetic metal; the wire is coiled on BB, DD, EE; C, outer soft iron tube; XXXX, soft iron plates connecting inner and outer soft iron tubes; KK, adjustable soft iron plug for adjustment.

instruments, that, according to the inventors, they may be used standing on the field magnets of a powerful dynamo machine without introducing any appreciable error. The scale is not proportional, the divisions getting wider apart as the deflection increases, nor is the range so wide as in the solenoid form. The adjustment in these instruments is effected by a screwed soft-iron plug at the bottom of the tubular magnet.

SHUNTS AND CIRCUIT RESISTANCE COILS.

Shunting the galvanometer.—A circuit placed between the poles of a galvanometer, for the purpose of reducing its sensibility in a certain known proportion, and to bring its deflections within the limits of the gradation, is called a shunt. To reduce the current to $\frac{1}{n}$ of its value, the resistance of the shunt S ought to be:

$$S = \frac{G}{n-1},$$

G being the resistance of the galvanometer.

Generally galvanometers are provided with a shunt box containing three shunts, which reduce its sensibility to the 10th, to the 100th, to the 1,000th, and of which the respective resistances are:

$$\frac{G}{9} \; ; \; \frac{G}{99} \; ; \; \frac{G}{999}.$$

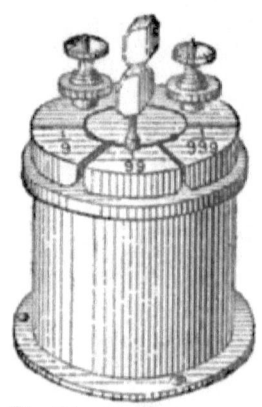

Fig. 28.—Galvanometer Shunt Box.

The shunts are enclosed in a separate box; Fig. 28 shows the arrangement generally used.

Multiplying power of a shunt.—The ratio of the current which traverses the galvanometer without a shunt to that which traverses the galvanometer with a shunt, the current traversing the whole circuit remaining the same. Calling this ratio m,

$$m = \frac{G+S}{S}.$$

Resistance of a shunted galvanometer.— Calling this G_1, then the law of derived currents gives

$$G_1 = \frac{GS}{G+S}.$$

When a galvanometer is shunted the value G_1 then comes into the calculations if no compensating resistance be used.

Compensating resistance.—When a galvanometer is shunted its resistance decreases, hence the current increases in strength; to bring back the current to its former strength resistance must be inserted in the circuit, which is called compensating resistance, and of which the value R_c is given by the formula

$$R_c = G\,\frac{n-1}{n} = \frac{G^2}{G+S}.$$

A galvanometer of resistance G, with its shunt and its compensating resistance, may be considered as a galvanometer of the same resistance, but smaller sensibility.

Constant of a galvanometer in French nomenclature is the deflection produced by one Daniell's cell in a circuit of which the total resistance is equal to one megohm. By shunting the galvanometer to the $\frac{1}{n}$, if r is the internal resistance of the Daniell's cell, G_1 that of the shunted galvanometer, R a resistance introduced into the circuit, such that

$$r + G_1 + R = \frac{1{,}000{,}000}{n},$$

the deflection of the galvanometer will be the constant.* In England the constant of a galvanometer means the number by which its indications must be multiplied to reduce them to ampères, milliampères, or microampères, as the case may be.

Maximum sensibility.—With a tangent galvanometer the maximum sensibility is at 45°. In measurements by the equal deflection method, the deflections must therefore be made about 45°. In half deflection methods the best angles are $35\tfrac{1}{2}°$ and 55°, for which the tangents, and consequently the strengths of the currents, are one the double of the other. In any kind of galvanometer it is as well to mark upon the instrument this angle of maximum sensibility.

Theorem of sensibility (*R. V. Picou*).—The relation between a physical action y, and the reading on a graduated scale x, may be written in the general form

$$y = A f(x),$$

A being a constant, and $f(x)$ a function which depends on the mathematical theory of the apparatus. The sensibility of the apparatus S is

---

* It would be more scientific and more practical to define the constant of a galvanometer as the deflection produced by a current of one microampère. The calculation of current strengths would be much simplified by this method, especially with tangent galvanometers, in which the deflections are proportional to the current strengths.

equal at each instant to the ratio of $f(x)$ to its differential coefficient $f'(x)$.

$$S = \frac{f(x)}{f'(x)} = \frac{y}{y'}.$$

The maximum point of sensibility is arrived at by considering the function S, and particularly by taking the value of x, which gives the value zero to the differential coefficient of the function S.

This theorem when applied to the tangent galvanometer indicates that its maximum sensibility is at 45°; for the sine galvanometer the sensibility increases infinitely, the maximum is at 90°.

Formula of merit of a galvanometer.—This is the resistance of a circuit through which one Daniell's element will produce unit deflection on the graduated scale of a galvanometer. The circuit is formed of one Daniell's cell of resistance r, a rheostat R, galvanometer G, and shunt S: a deflection of d divisions is obtained. The resistance of the shunted galvanometer is G_1.

$$G_1 = \frac{GS}{G+S}.$$

the multiplying power m of the shunt is: $m = \frac{S+G}{S}$.

Formula of merit $= md\ (r + R + G_1)$.

The formula of merit is larger as the galvanometer is more sensitive.

Circuit resistance coils.—Resistance coils, which are placed in the circuit of a calibrated voltmeter to increase the range of its indications; they have generally a resistance equal to 1, 2, 3 ... n times that of the voltmeter to which they belong. The readings made on the apparatus must therefore be multiplied by 2, 3, 4 ... $n+1$, in order to obtain the value of the quantity measured.

The sensibility diminishes in proportion to the number of coils introduced. These resistances, and the wire of the galvanometer, ought never to be allowed to heat, because the deflections would be reduced on account of the increase of resistance produced by heating.

Calibration of a galvanometer.—This operation consists in tracing out a gradation proportional to the strengths of the currents which pass through the galvanometer. With a tangent or sine galvanometer calibration is not required; it is only useful for apparatus for which the law of deflection is unknown. With any galvanometer of resistance G the operation is as follows: First of all shunts are prepared for the galvanometer of $\frac{1}{2}$, $\frac{1}{3}$, $\frac{1}{4}$, etc., and corresponding compensating resistances. A circuit is then formed composed of the galvanometer, a constant battery, and a resistance box. First of all a shunt $\frac{1}{2}$, and corresponding compensating resistance, is inserted. Sufficient resistance is then added to bring the deflection to a suitable value; for example, 1°. The shunt and the compensating resistance is then removed, the current passing through the galvanometer is thus doubled; the deflection obtained corresponds to a double strength. The shunt $\frac{1}{3}$, and its compensating resistance, is then inserted. The resistance box is then adjusted to bring the deflection back to the original value, say 1°. The shunt, and its compensating resistance, is then removed; the current is thus tripled; the deflection obtained corresponds to three times the strength, and so on. The deflections are marked on the galvanometer itself, or on a reduction table. To bring these deflections within the limits of the scale, a resistance may be inserted in the circuit, or the galvanometer may be shunted, or the battery may be shunted; but this last method makes the currents too large, and disturbs the constancy of the battery.

Absolute calibration of a galvanometer.— This operation consists in marking on the graduation of the instrument the current strengths in ampères corresponding to each deflection; it is especially used in commercial apparatus. The methods vary infinitely. One of them, based on electrolytic action, consists in causing a given current to pass through a decomposition cell, and the galvanometer to be calibrated for a certain length of time, t (seconds); keeping the deflection constant during the experiment, the quantity of electricity Q in coulombs which has passed through the decomposition cell, and the galvanometer deduced from the chemical action, enables us to calculate C from the relation:

$$C = \frac{Q}{t} \text{ ampères.}$$

Another method consists of introducing a perfectly fixed and known resistance R into the circuit of the galvanometer. The difference of potential between the two extremities of this resistance is measured by any suitable method, and C is deduced by Ohm's law. It is well to verify the calibration of a galvanometer often, as the calibration may change from moment to moment by the action of external or internal causes. This verification is made by the same means used for the original calibration.

Thickness and resistance of galvanometer wire; shape of the coils.—For very strong currents one

92   METHODS OF MEASUREMENT.

single turn of very thick wire is often used; for thermo-electric currents twenty to thirty of wire of one millimètre in diameter; the resistance is about a quarter of an ohm.

Galvanometers of high resistance (Thomson's, etc.) have from 5,000 to 10,000 ohms resistance; the diameter of the wire is not more than one-tenth or two-tenths of a millimètre, and its length may be as great as 4,000 mètres. Some galvanometers wound with German silver wire have as much as 50,000 ohms resistance; they give a deflection of 200 divisions of the scale with a single Daniell's cell, and 20 megohms in the circuit. The use of German silver is advantageous, especially in differential galvanometers, because of the small variation of resistance produced by changes of temperature. Resistance is always a disadvantage, but it is impossible to have a great number of turns of wire in a small space without a large resistance.

All contact between the wires must be avoided, as it would prevent the action of the whole part interposed between the points of contact. Bad insulation of the wire disturbs the true value of the shunts,

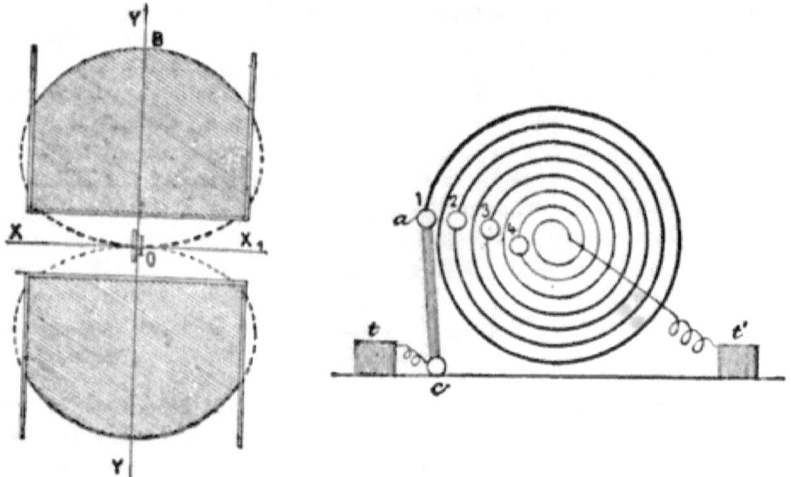

Fig. 29.—Form of the Coil in the Reflecting Galvanometer.   Fig. 30.—Graded Galvanometer.

and renders the instrument useless for exact measurements. Copper wire ought to be carefully covered with white silk, and well dried before it is coiled; after a few layers have been coiled the coil ought to be again dried, and steeped in pure paraffin.

The resistance of the wire as it is coiled ought to be frequently compared with its calculated resistance. For a given length and thickness of wire there is a special form of coil which gives the maximum effect. Sir William Thomson has calculated this form for the reflecting galvanometer; a transverse section of this form is given by the equation

$$x^2 = (a^2 y)^{\frac{2}{3}} - y^2.$$

x being the ordinate in the direction parallel to the axis of the coil, a the distance O B; O, the origin of the co-ordinates, the centre of the coil; Fig. 29 shows the theoretical curve and its practical form. Part of the wire must necessarily be removed from the centre to allow space for the magnet. The thickness of the wire ought not to be the same in all the layers.

The sectional area ought to increase proportionally to the diameter at each point in order to give the best results. In practice three or four different thicknesses answer the purpose. In Sir William Thomson's graded galvanometer, one, two, three, or four parts of the wire can be used according to need, the necessary connections being made by means of a key ac, which turns about the point c (Fig. 30).

Measurement of currents in C. G. S. units by the tangent galvanometer.—In the case of a galvanometer with a circular coil of which the needle is so short that the tangents of the angles of deflection are proportional to the current strengths:

r the radius of the coil in centimètres;
n number of turns of wire;
H horizontal component of terrestrial magnetism (in dynes);
C strength of current in C. G. S. units;
δ angle of deflection.
Then,

$$C = \frac{r}{2\pi n} \times H \tan \delta \text{ .G.S. units.}$$

And as 1 ampère $= \frac{1}{10}$ C.G.S. units,

$$C = \frac{r}{2\pi n} \times H \tan \delta \times 10 \text{ ampères.}$$

In England the ratio $\frac{r}{2\pi n}$ is called the constant of the galvanometer.*

---

* Here again we find ambiguity in consequence of badly defined expressions, as in France the constant of a galvanometer is the deflection produced by one Daniell's cell through a total resistance of one megohm.

Comparison of current strengths by the method of oscillations (*Latimer Clark*) is carried out by means of a galvanometer or galvanoscope with a single needle. The galvanometer coil is placed at right angles with the magnetic meridian, the needle is made to oscillate, and the number m of oscillations performed under the action of terrestrial magnetism in a given time (one minute, for example) is counted. A current of strength c is passed through the coil, and the number of oscillations n during the same time is noted.

The number of oscillations N is then counted with another current C. We have then the relation:

$$\frac{C}{c} = \frac{N^2 - m^2}{n^2 - m^2}.$$

If the horizontal component of the earth's magnetism H be known, the first two experiments are sufficient, and we have,

$$C = \frac{N^2 - m^2}{m^2} H.$$

Indirect measurement of current strength.—(1) *By Ohm's law.* The difference of potential E between two points of the circuit separated by a known resistance R is measured, and Ohm's law is applied:

$$C = \frac{E}{R}.$$

This method is particularly suitable for the measurement of very strong currents which do not permit of a galvanometer being placed directly in the circuit.

(2) *By the voltmeter.* The current to be measured is caused to pass through a voltmeter or decomposition cell for n seconds. The volume of gas is observed or the deposit is weighed, and the number of coulombs Q is calculated by the electro-chemical equivalents. The current strength C is then given by the formula,

$$C = \frac{Q}{n}.$$

II. ELECTRO-DYNAMOMETERS.

Electro-dynamometers depend on the mutual attractions and repulsions of currents. They give indications proportional to the square of the current strengths, and consequently independent of the direction of these currents. They are thus suitable for the measurement of alternating currents. As they contain no magnet, it is easy to make their indications independent of terrestrial magnetism.

Weber's electro-dynamometer is composed of a fixed coil and an interior concentric movable coil, of which the axis is at right angles to that of the fixed coil. It is supported by a wire bifilar suspension, the wires of which conduct the current, and their torsion balances the mutual action of the coils. The deflection is read by means of a lamp, scale, and mirror.

Joule's electro-dynamometer. — The movable coil is suspended from a scale beam. It is horizontal, and is free to move vertically. The fixed coil is below it. The planes of the turns of the two coils are parallel. The force exerted between the two coils is measured by the weight, which must be added to or taken from the scale pan suspended to the other end of the beam, and balancing the movable coil.

Siemens and Halske's electro-dynamometer, intended for practical work. The action of the current is balanced by the torsion of a bifilar suspension, hair or spiral spring. It is composed of a fixed coil, and a movable coil outside the fixed coil, and having only one single turn of wire. The directive action of the earth upon the movable coil may then be entirely neglected; that of the fixed coil is proportional to the number of turns of wire. At each measurement the two coils are brought back into a position at right angles to each other, and the angle of torsion, which may be as much as 270°, measures the current. The sensibility increases with the current strength, since the torsions are proportional to the squares of the current strengths. The most complete form of this instrument has two fixed coils. One of the coils made of a thick wire is used for currents of from ten to sixty ampères; the other coil for currents of from one-half to ten ampères.

III. VOLTMETERS.

Up to the present time the measurement of current strengths by the voltmeter has been but little applied. This method is based on Faraday's law. A constant current of unknown strength C is caused to pass through a voltmeter or decomposition cell for t seconds. The volume of gas disengaged is measured, or the deposit produced by the passage of the current is weighed by means of the electro-chemical equivalents. The quantity of electricity Q in coulombs which has passed through the voltmeter is calculated. The value of C is then deduced from the formula:

$$C = \frac{Q}{t}.$$

This method is principally employed for the calibration of galvanometers. (*See* Measurement of electrical quantities.)

MEASUREMENT OF RESISTANCES

The methods of measuring resistances are very numerous, and vary with the kind of instrument at hand, the accuracy which it is desired to obtain, and the nature of the resistance to be measured. We will point out here the most simple and most commonly used methods.

RESISTANCE OF CONDUCTORS.

Substitution method.—A constant battery, a galvanometer G and the resistance to be measured x, are arranged in circuit. The deflection of the galvanometer is noted, and a resistance box is substituted for x. The box is unplugged until the deflection is the same as at first.

If R be the resistance unplugged, we have $x = R$. The accuracy depends on the sensibility of the galvanometer, the accuracy of the box, and the constancy of the battery.

By addition to a known circuit.—A circuit of total resistance R, made up of a battery, a galvanometer, and a resistance box, gives a deflection δ. The resistance to be measured x is introduced, and the deflection becomes δ'. We have then,

$$\frac{\delta}{\delta'} = \frac{R + x}{R},$$

Whence

$$x = \frac{\delta - \delta'}{\delta} R.$$

δ and δ' are not the angles, but the current strengths, corresponding to the deflections. They may be expressed in any arbitrary units.

This method requires a calibrated galvanometer and a constant battery.

Wheatstone's bridge.—The most convenient form for ordinary resistances is the Post-Office bridge box. When suitable resistances

WHEATSTONE'S BRIDGE.

have been unplugged between AB and BC, the battery key is pressed down, resistances are unplugged in the box which nearly correspond to equilibrium, and the left-hand key, which corresponds to the galvanometer, is pressed down. The plugs are then taken out or put in, as may be required, until the needle of the galvanometer remains at zero. If the galvanometer is very sensitive, it must at first be shunted, and very quick blows be given on the keys, so as not to risk breaking the suspending filament. When equilibrium is obtained, the shunt may be removed from the galvanometer, and the circuit kept closed for a longer time. The figure below shows the arrangement of the circuits; K_1 is the battery key (the right-hand key in the box), K_2 the galvanometer key (the left-hand key in the box), x the resistance to be measured, and AD the resistance box. According to the resistance of the galvanometer and the value of the resistances to be measured, different arrangements may be adopted for the ratio of the arms of the bridge, or the galvanometer may be placed where the battery usually is, and the battery in place of the galvanometer.

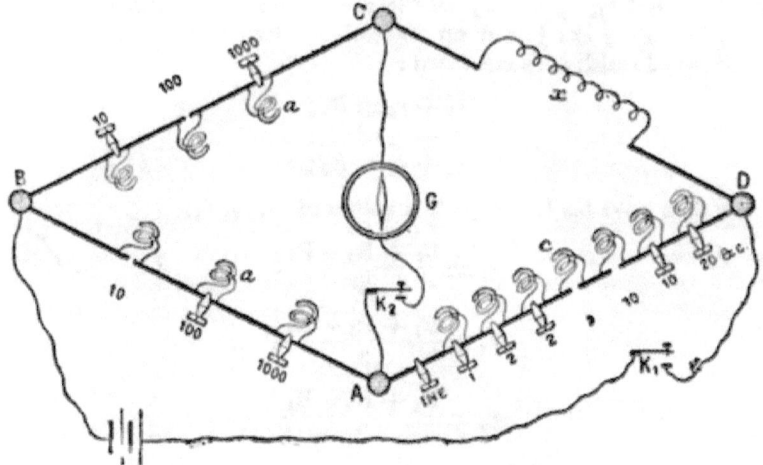

Fig. 31.—Diagram of Wheatstone's Bridge.

Resistance of galvanometer for maximum sensibility (Schwendler).—The resistance G of the galvanometer which gives the greatest sensibility for a given arrangement of the bridge is:

$$G = \frac{(a+b)(c+x)}{a+b+c+x}.$$

H

This formula allows a suitable galvanometer to be chosen when we know the sort of resistances to be measured.

Measurement of the resistance of a conductor which is put to earth.—The point D of the bridge is connected to earth; one of the ends of the resistance x is connected to the point C, the other to earth. One of the poles of the battery is also connected to earth. Generally two different values are found, R' and R'', according to whether a positive or negative current has been used, because of the influence of the earth couple on the resistance to be measured. If the readings are taken quickly, we may take it that

$$x = \frac{R' + R''}{2}.$$

The resistance x then includes that of the conductor and the sum of the earth resistances.

Resistance of overhead lines.—*When three lines are available.*—Let r_1, r_2, and r_3 be the resistances of the three lines to be measured. They are joined up successively, two by two, in circuit, and the combined resistances measured:

$$r_1 + r_2 = R_1;$$
$$r_1 + r_3 = R_2;$$
$$r_2 + r_3 = R_3.$$

We then have for the respective values of r_1, r_2, r_3:

$$r_1 = \frac{R_1 + R_2 - R_3}{2};$$

$$r_2 = \frac{R_1 + R_3 - R_2}{2};$$

$$r_3 = \frac{R_2 + R_3 - R_1}{2}.$$

Ayrton and Perry's ohmmeter.—Founded on Ohm's law and the measurement of R by the ratio $\frac{E}{C}$. Two coils fixed at right angles act on one and the same needle. One of the coils, which is wound with thick wire, is placed in the main circuit; the other, wound with fine wire, as a shunt between the two extremities of the resistance to be measured. By making the coils and the needle of suitable proportions, the deflections are proportional to the resistances, and the measurement is made by a reading on the graduated dial. The ohmmeter does away with the necessity for a galvanometer and a resistance box, and enables a conductor traversed by currents to be measured *when hot* without stopping the machines.

J. Carpentier's proportional galvanometer.—
Two coils arranged at right angles, and wound with the same number of turns of wire. In the centre a small needle and mirror. These coils are connected up in parallel arc. A known resistance being placed in the circuit of one of them, and the resistance x to be measured in the circuit of the other, the deflection of the needle read by means of the mirror shows the resistance directly.

Measurement of the specific conductivity of a conductor.*—
Is used especially for copper, of which the true conductivity at 0° is represented by 1. In order to find the specific conductivity of a given sample, the real resistance of this sample is measured, and the resistance from its dimensions which it ought to have at the temperature of the experiment, if it were pure, is calculated. Let R_m be the measured resistance and R_c the calculated resistance ; then,

$$Conductivity = \frac{R_c}{R_m}.$$

The number given by the formula is always smaller than 1.

Measurement of very large resistances.—
(1) A current from a battery of e. m. f. E is passed through a resistance x so great that the resistance of the battery and galvanometer may be neglected in comparison with it. A deflection δ is obtained so that

$$\delta = \frac{E}{x}.$$

If the *constant* of the galvanometer is d, we have:

$$x = E \frac{d}{\delta} \text{ megohms.}$$

(2) n elements of e. m. f. E, the resistance to be measured x, and the galvanometer G, are arranged in circuit. A certain deflection δ is

---

* Conductivities are often expressed by taking the conductivity of pure copper at 100. In this case the number which represents the conductivity relatively to pure copper taken as unity, must be multiplied by 100.

obtained. The same deflection is then obtained with n' elements, and a rheostat, the galvanometer being shunted to the $\frac{1}{m}$.

Then:
$$x = \frac{n}{n'} mR.$$

If $\frac{n}{n'} = 100$ and $m = 1,000$, $x = 100,000$ R.

This method is used for the measurement of the insulation of telegraph lines.

Measurement of very low resistances.—With very low resistances, bad contact affects the result in the usual methods. For these measurements *Thomson's bridge* is used. In the figure, x is the

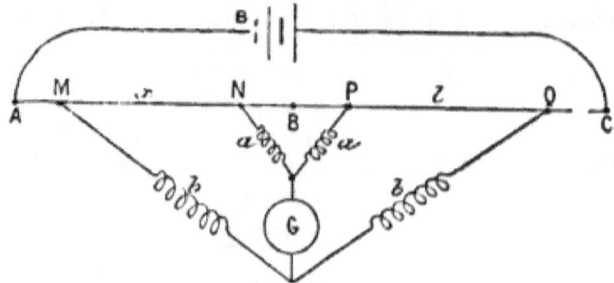

Fig. 32.—Measurement of very Low Resistances.

resistance to be measured between the points M and N; l is a graduated wire of which the resistance is known, a and a two equal resistances, b and b two equal resistances, G is a sensitive galvanometer, B the battery. The points P and Q are shifted until the galvanometer comes to zero, then

$$x = l.$$

RESISTANCE OF GALVANOMETERS.

Half deflection method.— A constant battery of small resistance is used. A resistance box, battery, and galvanometer of resistance G are placed in the same circuit. A resistance R is introduced into the circuit, and this resistance is increased up to a value R_1, such that the current strength is reduced to one-half; we have then,

$$G = R_1 - 2R.$$

This method requires a calibrated galvanometer.

Equal deflection method.—With a low resistance constant battery and non-calibrated galvanometer. The galvanometer G, the battery E, and shunt S, and resistance box R are arranged as shown

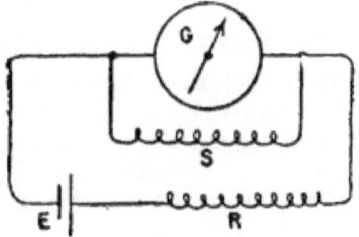

Fig. 33.—Equal Deflection Method.

in the figure. The resistance R gives a certain deflection of the galvanometer; the shunt is removed, and the resistance R is increased up to a value R_1, such that the deflection becomes the same as in the first case. Then:

$$G = S\left(\frac{R_1 - R}{R}\right).$$

Sir William Thomson's method, which is independent

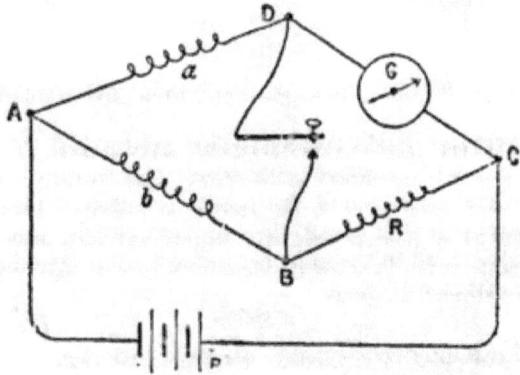

Fig. 34.—Thomson's Method.

of the resistance of the battery. The Wheatstone bridge is arranged as shown in the figure. The resistance box R is then varied until the deflection of G does not change, when the short circuit key between B and D is closed, then

$$G = R \frac{a}{b}.$$

INTERNAL RESISTANCE OF BATTERIES.

Half deflection method.—Applicable to batteries without sensible polarisation. The battery of unknown resistance r, a galvanometer of resistance G, and a resistance box are arranged in one circuit; a resistance R is unplugged, giving a deflection a, this resistance is then diminished until the deflection is $2a$. If R' is this second resistance,

$$r = R - (2R' + G).$$

a and $2a$ must be replaced by their corresponding sines or tangents, according to the kind of galvanometer used.

Sir William Thomson's method.—Applicable to batteries without sensible polarisation. The battery of resistance r, galvanometer G, and resistance box are placed in circuit; such a resistance R is then unplugged as will give a deflection easy to be read, and at the point of good sensibility of the galvanometer. A shunt of resistance S is then put between the poles of the battery, and the galvanometer is brought back to the same deflection by diminishing the resistance in the box to a value R_1, then

$$r = S \frac{R - R_1}{R_1 + G}.$$

This method may be used with a non-calibrated galvanometer.

Differential galvanometer method (*Latimer Clark*).—Galvanometer with a short thick wire. The current passes through one of the coils of resistance G, the needle is deflected through an angle a, and the current is then passed through both circuits, and the galvanometer is brought back to the same deflection by the introduction into the circuit of a resistance R, then

$$r = R.$$

This method can only be applied to constant batteries.

Measurement of internal resistance of batteries when an even number of absolutely identical elements is at hand.—They are put up in two series groups, with the same number of cells in each group. These groups are then connected in opposition. The e. m. fs. balance each other, and then the total resistance is measured in the same way as that of an ordinary conductor by any known method. (Substitution, Wheatstone's bridge, etc.)

Method by means of the electrometer, condenser, or galvanometer of very high resistance.
—The electrometer, the condenser, or the high resistance galvanometer is connected to the two poles of the battery, which is otherwise on open circuit, so as to measure the difference of potential either by a discharge method or by a direct reading; the system is then shunted by a known resistance until the difference of potential is reduced to one half, the resistance of the shunt is then equal to that of the battery. Only applicable to constant batteries.

Mance's method, one of the best, as it only requires the battery to be constant during the short interval during which the key is closed.

A Wheatstone bridge being arranged, as shewn in the figure, with a

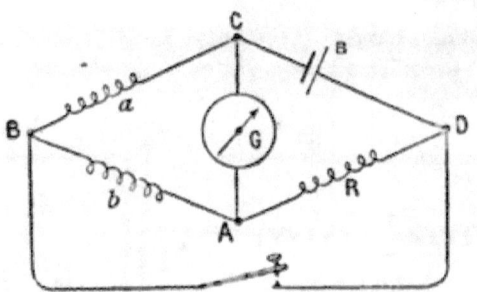

Fig. 35.—Mance's Method.

short circuit key between B and D, the arm AD is adjusted, until on pressing down the key the deflection of the galvanometer does not change, then

$$r = R \frac{a}{b}.$$

If the arms a and b are equal, the formula is simplified, and becomes

$$r = R.$$

Siemens' method requires a continuous rheostat, or a sliding contact resistance box. Two points B and B_1, in the resistance AC, are found such that the deflection of G does not change; in this case

$$r = G + b - a.$$

The galvanometer ought to be of smaller resistance than the battery, and

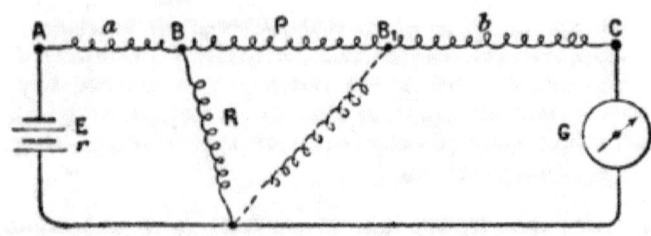

Fig. 36.—Siemens' Method.

sufficiently sensitive to allow R to be made fairly small without reducing the deflection too much.

Munro's method.—B is a battery of which the internal r is to be measured, C a condenser from $\frac{1}{3}$ to 1 microfarad, S a shunt, and

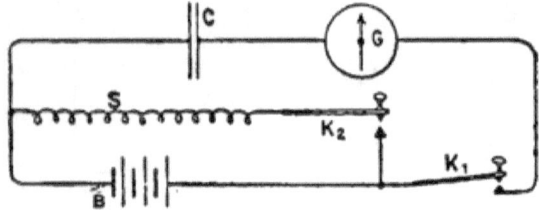

Fig. 37.—Munro's Method.

G the galvanometer; K_1 and K_2 keys. K_1 is pressed down, and the deflection of the galvanometer d_1 is observed; keeping the key K_1 pressed down, K_2 is pressed down, and the deflection d_2 in the opposite direction is observed; then

$$r = S \frac{d_2}{d_1 - d_2}.$$

This method is one of the best in practice, as it is applicable to all batteries.

INSULATION OF OVERHEAD LINES.

General measurement.—A tangent galvanometer of resistance G, a battery, and a fixed resistance of say 1,000 ohms, are put up in circuit; the deflection δ is observed; this is taking the constant of the galvanometer; then the resistance box is removed, and the free pole of the battery is connected to earth, and one of the ends of the line to the galvanometer, the other end remaining insulated. A second deflection δ' is thus obtained; the insulation resistance of the line R_i is then

$$R_i = 1000 \times \frac{\delta}{\delta'}.$$

In order to make the influence of the earth current negligible it is as well to use from thirty to forty Daniell elements in series.

Insulation per mile.—If the line is n miles long, the insulation per mile is $R_i n$. In good conditions the insulation per mile ought not to be less than 300,000 ohms. (*See* Fourth Part.) This method of calculating the insulation per mile is not very correct, because it supposes that the leakage is identical at every point of the line. The measurement may be made more correctly by taking into account the resistance G of the galvanometer, and resistance r of the battery;* the formula is then

$$R_i = (1000 + r + G)\frac{\delta}{\delta'} - (r + G).$$

When a great number of lines have to be measured at once it is useful to arrange a double entry table, in which is noted the insulation resistance for all values of δ and δ'.

MEASUREMENT OF POTENTIALS AND ELECTROMOTIVE FORCES.

The difference of potential between two points of an electrified circuit is measured directly or indirectly. The direct measurement is obtained by a special class instrument, the electrometer. There are a great number of indirect methods which enable these measurements to be made. All galvanometers, for example, which in reality only measure current strength, may also be used for the measurement of potentials or electromotive forces.

---

* r is the resistance of the whole battery, not that of one element.

ELECTROMETERS.

Electrometers belong to two classes, according as they are based upon (1) *electrostatic* actions, or (2) *electro-capillary* actions. When they only *show* differences of potential they act as *electroscopes;* when they *measure* these differences they are *electrometers*.

Electroscopes.—The best known is the gold-leaf electroscope. It is composed of two strips of gold leaf from 8 to 10 centimètres long by 2 broad, suspended in a glass globe by a rod of metal terminated by a plate of brass, which, when it is electrified, causes the gold leaves to diverge; in *Bohnenberger's electroscope* there is only one gold leaf suspended between two bodies; one charged with positive and the other with negative electricity. These instruments are very sensitive, but are not much used for the purposes of measurement.

Repulsion electrometers.—*Cavendish's* (1771-1781) and *Lane's* (1772) may be used for rough measurements; the first true electrometer is *Coulomb's torsion balance* (1785). In *Milner's* and *Peltier's* electrometers the torsion thread is replaced by a magnetised needle which produces the directive force; the same device is used in *Kohlrausche's* apparatus. These instruments are not much used now, being replaced by Sir W. Thomson's *absolute* and *quadrant* electrometers.

Sir William Thomson's absolute electrometer. —Based upon the attraction of two electrified discs arranged parallel to each other. One of the discs of known dimensions is surrounded by a guard ring, which causes the charge upon the disc to be uniformly distributed as if it had no edges. One of the discs 's suspended by springs; a micrometer screw is so regulated that the disc remains suspended a little above the guard ring when no part of the apparatus is electrified.

Idiostatic method.—The two plates are connected with the two bodies, the difference of potential between which is to be measured. The movable plate is then raised up until it takes up its original position, which is observed by means of a stretched hair and two fixed marks. At this moment the force of the springs and the attraction between the two discs balance. Calling V the potential of one of the plates, and V' that of the other, the difference of potential is given by the formula

$$V - V' = D \sqrt{\frac{8\pi F}{A}}.$$

D, distance between the plates.

F, electrical attraction equal to the effort of the springs which balance it.

A, mean area between the surface of the suspended disc and the opening of the guard ring.

This method of using the absolute electrometer is an idiostatic method, because no external charge is introduced. It is necessary that the exact distance D between the two discs should be known.

Heterostatic method.—In this method the two plates are insulated; the upper one is charged to a high and constant potential. The constancy is *verified* by the aid of an accessory electrometer or gauge, and this constancy is *maintained* by means of a *replenisher*, which re-charges the disc. The lower plate is alternately connected to the earth and to the body of which the potential is to be measured. The difference of attraction in the two cases gives the difference of potential between the body and the earth, that is to say, the potential of the body.

The formula then becomes,

$$V - V' = (D - D') \sqrt{\frac{8\pi F}{A}}.$$

$V - V'$ is the difference of potential between the earth and the electrified body; $D - D'$ the difference of the readings of the screw of the lower plate, which may be observed with perfect accuracy without introducing the absolute distance between the plates; very great correctness is thus obtained.

Sir William Thomson's quadrant electrometer

is composed of a needle in the shape of a figure 8 suspended by means of a bifilar suspension between four horizontal metallic quadrants, which are electrically connected together diagonally two by two. The needle is kept positively charged by means of a Leyden jar, and its charge is kept constant. (*See* Heterostatic method, Gauge, and Replenisher.) One of the pairs of quadrants is connected to earth (potential $= 0$ by definition), the other is connected to the body, of which the potential is to be measured. The deflection is a function of the difference of potential. According to the form of the needle, and the relative dimensions of the quadrants and the needle, the deflections measured in degrees are proportional to the differences of potential up to $3°$ in general, and up to $10°$ when the instrument is well constructed, and is used under good conditions. The readings are made on a curved scale by means of a lamp and mirror. The best type, besides the gauge and replenisher, is furnished with an arrangement by which the directive force can be varied, and by which it may be observed whether this force remains constant after it has been adjusted; and an *induction plate*, which diminishes the sensitiveness of the instrument. In measuring high potentials the induction plate is connected to the electrified body. The potential induced by this plate, which is small and far from the quadrants, is thus measured.

Law of deflection of the quadrant electrometer (*Clerk-Maxwell*).

$$M = K (A - B) [C - \frac{1}{2}(A + B)].$$

M, moment of the couple which turns the needle.
A and B, respective potentials of the two pairs of quadrants.
C, potential of the needle.
K, constant of the instrument.

If A and B are equal potentials of contrary signs, the electrometer becomes symmetrical, and the equation is reduced to

$$M = K (A - B) C.$$

Mascart's symmetrical electrometer.—A simplified form of Thomson's quadrant electrometer. It is used by the heterostatic method. The needle is connected to the body of which the potential is to be measured; each pair of quadrants to one of the poles of a chloride of silver battery of twenty to forty elements, of which the middle is connected to earth in order to give equal charges of contrary signs to the quadrants; the apparatus is then symmetrical. This method is used for observing the atmospheric electricity at the observatory of Montsouris. In the arrangement adopted by M. Mascart the moment M which turns the needle is nothing when

$$C = \frac{1}{2}(A + B).$$

Lippmann's capillary electrometer.—A very sensitive instrument intended for the measurement of very small e. m. fs., based on the variations which the capillary depression of mercury undergoes under the influence of an e. m. f. In the latest form designed by M. Lippmann the depression caused by the e. m. f. is balanced by a pressure exerted on the mercury by means of a pneumatic arrangement. The height of the mercury is read by means of a microscope, and is brought back to the same point at each experiment. The value of the pressure exerted measured by a mercury pressure-gauge gives the e. m. f. The instrument is most sensitive between zero and one half Daniell. It will show $\frac{1}{10000}$ of a Daniell. Its indications are very quick, and enable

the variations of an electrical phenomenon of short duration and varying with time to be observed (loss of charge of a condenser of a secondary battery, etc.). Its great sensitiveness enables it to be used in all zero methods (Wheatstone's bridge, etc.).

Debrun's capillary electrometer.—The essential part is a capillary tube one millimètre in diameter, arranged almost horizontally, in which the mercury is displaced under electrical action; it is sensitive enough to show the $\frac{1}{10000}$ of a volt. It is graduated by means of zinc-cadmium elements, of which the e. m. f. is ·281 volt.

Ayrton and Perry's cylindrical spring electrometer. — For the measurement of potentials above 500 volts. On the same principle as the quadrant electrometer. In this instrument the quadrants are quarters of an elongated cylinder, and the needle, two cylindrical plates attached to a vertical axis; a spiral spring balances the torsion produced by the attractions due to the charges of the two pairs of quadrants which are joined to the two points, the difference of potential between which is to be measured. The torsion of the spring measures the difference of potential. It is used by the idiostatic method, by joining the movable cylinder to one of the pairs of quadrants. It enables the e. m. f. of alternating current machines to be measured, which cannot easily be done by electro-dynamometers, because of self-induction. The instrument is portable and fairly dead-beat, the movable cylinder having a small moment of inertia.

INDIRECT MEASUREMENT OF DIFFERENCES OF POTENTIAL.

We will go over the principal *indirect* methods of measuring the difference of potential D between two points A and B.

Graduated galvanometers or voltmeters. — If we have a galvanometer, of which the function which connects current strengths to the deflections is known, and of so high a resistance, that if it is connected as a shunt between the points A and B, so small a current only passes through it that the flow of the current in the rest of the system is not sensibly altered; Ohm's law enables us to deduce at each instant the difference of potential between the points A and B from the strength of the current which passes through the galvanometer. In practice voltmeters are constructed with a resistance of several thousand ohms, and they are graduated directly in volts. Measurement is thus reduced to a direct reading. The galvanometers of *Sir William Thomson*,

Marcel Deprez, *Ayrton* and *Perry*, etc., are thus constructed, and are used in practical measurements of machines, motors, and lamps.

Opposition method.—A sensitive galvanometer, and n elements of e. m. f. E in series, are arranged between the points A and b, so as to send a current in the opposite direction to that which would flow through a conductor connected to the two points A and B; n is then varied until the galvanometer comes to zero, or the deflection changes signs according as there are n or $n + 1$ elements. Then,

$$nE < D < (n + 1) E.$$

The error cannot be greater than $\frac{E}{2}$, which is generally sufficient in practice.

Partial opposition method.—Two resistance boxes, R and R' (Fig. 38), of so high a resistance as not sensibly to alter the difference

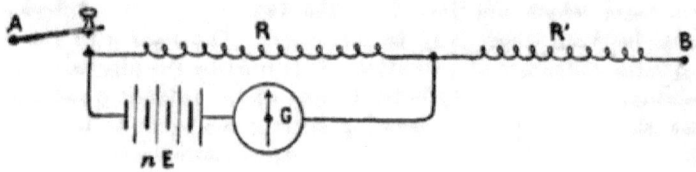

Fig. 38.—Partial Opposition Method.

of potential, are placed between A and B. A galvanometer G. and a battery nE are arranged as shown in the diagram. R and R' are then varied until the galvanometer comes to zero. Then,

$$D = nE \frac{R + R'}{R}.$$

The opposition methods have the advantage of not polarising the standard battery, thus giving more exact measurements, the galvanometer only acting as a galvanoscope.

Condenser method.—This method is identical with that employed in measuring the e. m. f. of batteries, which we will describe later on.

E. M. F. OF BATTERIES.

The e. m. f. of a battery is equal to the difference of potential between its poles when the battery is on open circuit. For want of a standard of e. m. f., the e. m. f. of a battery is measured by comparison with that of another battery taken as a unit; it is then expressed in practical units or volts by multiplying the result by the e. m. f. of the standard which has been used.

Equal resistance method.—A battery of internal resistance r, of which the e. m. f. is to be measured, a galvanometer G and a resistance box R, are arranged in circuit. R is varied so as to obtain a deflection within the limits of the graduation of G, the strength of the current is then C. The battery is replaced by the standard, and R is so varied as to make the total resistance of the circuit the same as in the first case; the current strength is then C', whence,

$$\frac{E}{E'} = \frac{C}{C'}.$$

According to the nature of the galvanometer C and C' are expressed by the tangents or the sines of the deflections.

When the resistance of the galvanometer, together with the resistance R, is very large compared with the internal resistance of the batteries to be measured, there is no necessity to equalise the total resistance in the two experiments. This is the case, for example, when the total resistance of the galvanometer and the box exceeds from 20,000 to 25,000 ohms.

Equal deflection method is used when the galvanometer is not calibrated; the standard E, galvanometer G, and the box are arranged in circuit; the box is adjusted so as to have a convenient reading on the galvanometer. Let R be the total resistance; the standard is replaced by the battery to be measured E', and the galvanometer is brought back to the same deflection; the total new resistance is R, whence,

$$\frac{E}{E'} = \frac{R}{R'}.$$

The internal resistance of the elements may be neglected in comparison with that of the galvanometer G, and the resistances R and R' introduced into the circuit when these resistances are large; the formula then becomes,

$$\frac{E}{E'} = \frac{R + G}{R' + G}.$$

Wiedemann's method.—Let E be the e. m. f. of the standard battery and E' that of the battery to be measured. The two batteries, a galvanometer, and the resistance box are placed in circuit. Let d be the deflection on the tangent galvanometer due to the sum of the two e. m. fs. The weaker of the two batteries is then reversed so as only to have the current due to the difference of the e. m. fs. Let d_1 be the new deflection; then,

$$\frac{E}{E'} = \frac{d + d_1}{d - d_1}.$$

Wheatstone's method.—A battery of e. m. f. E is introduced into the circuit of a galvanometer G and resistance box R, a certain deflection a is then obtained, then a new resistance p is introduced so as to obtain a smaller deflection B. The battery is then withdrawn, a battery of e. m. f. E' is substituted for it, the resistance box is adjusted so as to bring the deflection back to the value a, obtained by the first battery. A resistance ρ' is then added so as to bring the deflection back to the value B; then,

$$\frac{E}{E'} = \frac{\rho}{\rho'}.$$

This method does not require a calibrated galvanometer, and is independent of the internal resistance of the elements.

Lacoine's method.—Two batteries of e. m. f. E and E' are arranged in series, and a galvanometer G (Fig. 39) arranged as a shunt

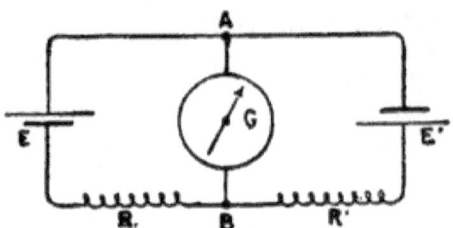

Fig. 39.—Lacoine's Method.

between the points A and B. Between E and the point B a certain resistance R is introduced, and the resistance R' is adjusted until the galvanometer comes to zero; then,

$$\frac{E}{E'} = \frac{R}{R'},$$

supposing the resistance of the batteries to be negligible as compared with R and R'. When these resistances are so large that they must be taken into account the method is thus modified:

A first experiment is made with the resistances R and R′, then R is changed by giving it a smaller value R_1, and R′ is adjusted so as to bring the galvanometer back to zero, the new value of R′ is R'_1; then,

$$\frac{E}{E'} = \frac{R - R_1}{R' - R'_1}.$$

If the internal resistance of the elements is known this second operation is not needed. Calling the internal resistances r and r' the formula then becomes,

$$\frac{E}{E'} = \frac{R + R}{R' + r'}$$

Poggendorff's method.—A zero method. The batteries of

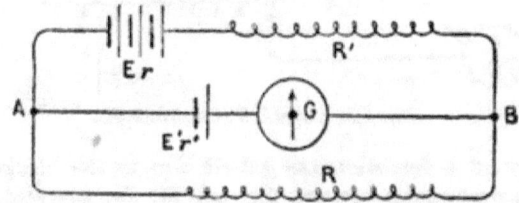

Fig. 40.—Poggendorff's Method.

e. m. f. E and E′ are arranged as in Fig. 40; R and R′ are adjusted until there is equilibrium; then,

$$\frac{E}{E'} = \frac{R + R' + r}{R}.$$

In this method of measurement the battery E′ produces no current, and does not polarise. The battery E ought to be constant, and be formed of a sufficient number of, say, Daniell elements, that E may be greater than E′. The internal resistance r of the battery E comes into the above equation. It may be eliminated by making two experiments, the first with the resistances R and R′, and secondly with smaller resistances R_1 and R'_1; the formula then becomes,

$$\frac{E}{E'} = \frac{(R - R_1) + (R' - R'_1)}{R - R_1}.$$

Clark's potentiometer.—Requires two galvanometers and three batteries, the standard, the battery to be measured, and an auxiliary battery. It has the advantage that the standard and the battery to be measured are compared under the same conditions, no current passing through either of them. Thus errors produced by polarisation which are introduced into most other methods, are avoided.

In the above diagram R is a coil of bare wire, made of platinum and iridium alloy of 40 ohms resistance, making 100 turns round an ebonite cylinder, turning on an axle like a Wheatstone's rheostat. The two ends of the wire are attached to the extremities A and B, which serve as pivots.

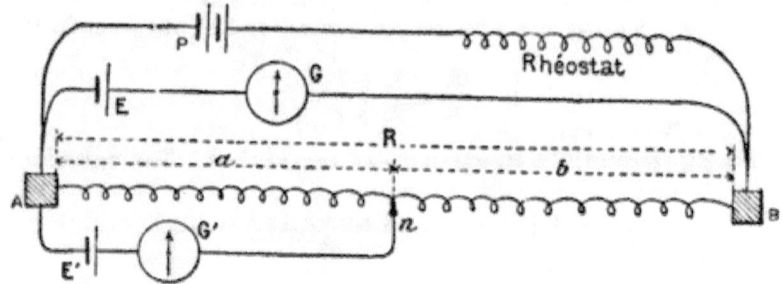

Fig. 41.—Clark's Potentiometer.

P is a battery of a few elements joined also to the blocks A and B, which sends a continuous current through R; the rheostat enables the total resistance in this circuit to be varied. The standard is at E joined to the points A and B with a galvanometer interposed at G, which must be brought to zero; this may be easily done by varying the rheostat. The battery E' is joined to the point A by one of its poles, the other pole is joined to a second galvanometer G', and a contact n, which slides on the resistance. The point of contact n is moved along the resistance R until the galvanometer G' comes to zero. Calling the two parts of the resistance R on each side of the contact a and b, when G' is at zero, we have,

$$\frac{E}{E'} = \frac{a+b}{a} = \frac{R}{a}.$$

The resistance R being graduated, the ratio is read directly on the scale. The error in this method is less than the $\frac{1}{1.000.000}$th of a volt.

When the battery to be measured is stronger than the standard they are interchanged, the standard is put at E' the battery at E, and the experiment is made as before. It is only necessary to substitute the letters one for another in the formula when the elements have been exchanged. Prof. Adams has justly remarked that the galvanometer G' is useless, because the galvanometer G being at zero for a certain value of the rheostat, its equilibrium would be disturbed if the slider n were in any other position than that for which the galvanometer G comes to zero.

DIFFERENCES OF POTENTIAL.

Law's method.—One and the same condenser is charged successively by the two batteries which are to be compared; the ratio of the charges shows the ratio of the e. m. fs. The charges are measured by means of a balistic galvanometer, or a galvanometer with a suspended needle. The angle of impulse is almost proportional to the e. m. f., or, more exactly, the e. m. f. is proportional to the sine of half the angle of impulse, but when the deflection is not too great the e. m. f. is practically proportional to the angle. By shunting the galvanometer the e. m. f. of a whole battery may be obtained in terms of a single standard element. In this case the impulse produced by the standard element is observed, and the galvanometer is shunted until the whole battery gives the same impulse; then, if S be the resistance of the shunt,

$$E' = \frac{G+S}{S} E;$$

If the angles of impulse are not equal, calling that produced by the standard δ, and that produced by the battery δ', we have

$$E' = \frac{\delta'}{\delta} \cdot \frac{G+S}{S} E;$$

or more correctly,

$$E' = \frac{\sin \frac{1}{2} \delta'}{\sin \frac{1}{2} \delta} \cdot \frac{G+S}{S} E.$$

Correction for the resistance of the air.—After having observed the first impulse the scale is carefully watched, and the point to which the spot of light comes on the second swing is also observed. One quarter of the difference between the two readings is added to the first reading in order to correct for the resistance of the air.

Opposition method.—n elements of known e. m. f. E are opposed to n' elements of unknown e. m. f., E' interposing a galvanometer. n and n' are then varied until the galvanometer comes to zero; then,

$$nE = n'E'.$$

If n elements give a deflection δ on a galvanometer on one side of zero, and $n+1$ elements a deflection δ' on the opposite side, then,

$$n \frac{\delta}{\delta + \delta'} E = n'E'.$$

From which the value of E' may be found.

MEASUREMENT OF ELECTRICAL QUANTITY.

Faraday's law.—If C be the strength of the current, t the time during which it is passing, the quantity of electricity Q given by the current during the time t is

$$Q = Ct.$$

Taking C in ampères and t in seconds, we get the **value Q** in coulombs. This is the method used for calculating the elements of dynamos which are to be used for electro-metallurgic operations; it is an indirect method. Direct methods, based on the chemical or mechanical actions of the current, are carried out by means of *voltmeters, coulomb meters*, or *electricity meters.*

Voltmeters. — On account of the tendency of the gases to dissolve in acidulated water and other secondary phenomena, the gas voltmeter is not very correct, and is but very little used, and we only notice it for the sake of completeness.

Electrolytic cells.—The metal deposited by a current in a given time is weighed, and the number of coulombs is deduced by the electro-chemical equivalent. The solutions most used are sulphate of copper, sulphate of zinc, and nitrate of silver.

M. Mascart has made some experiments with a 15 per cent. solution of nitrate of silver and a 10 per cent. solution of sulphate of copper.

Edison's meters.—Sulphate of copper was used in the first; in the later ones, sulphate of zinc. It is set up in a derived circuit, so that only $\frac{1}{100}$th or $\frac{1}{1000}$th of the total current passes through it. The zinc solution contains 90 parts by weight of pure sulphate of zinc dissolved in 100 parts of distilled water; its density at 18° ought to be 1·33 (*Francis Jehl*). The zinc plates are weighed once a month, and the number of coulombs is deduced from this weight by remembering that

1 *ampère hour* deposits 1,228 milligrammes of zinc.

In another instrument of Edison's weighing is dispensed with. An automatic arrangement causes the plates to tip over as soon as they have gained a weight exceeding a certain weight; the connections are changed, the plate which was the cathode becomes the anode, and reciprocally until the apparatus tips over the other way. A counter registers the number of movements produced during a given time. The total number of coulombs is deduced from this number by a very simple calculation. The apparatus is rather complicated.

Edison's and Ayrton and Perry's coulomb meter.
—The principle of this instrument is to use an electromotor so arranged that its speed is proportional to the strength of the current passing through it. This motor turns a fan immersed in liquid. The resistance to the movement being thus also proportional to the speed, if the number of revolutions performed by the apparatus during a given time be registered, the number of coulombs which have passed through it can be found. It has not yet been used in practice.

Vernon-Boys' integrating meter.—This apparatus is an integrator which gives directly for any time t by a simple reading:

$$\int C dt.$$

It has not yet been used in practice.

MEASUREMENT OF CAPACITY.

Capacities are measured by comparing them to those of standards, which in general vary between $\frac{1}{3}$ and 1 microfarad. In practice these measurements are only applied to submarine cables, so we will only point out here a general method, reserving the explanation of special methods for the sections on cable measurements.

Electrostatic capacity of condensers.—A standard condenser of known capacity c is charged by means of a battery of given e. m. f., and discharged through a balistic galvanometer. Let a be the deflection. A condenser of capacity c_1 is then charged with the same battery. It is discharged, and a second deflection a_1 is obtained; then,

$$\frac{c}{c_1} = \frac{a}{a_1}.$$

When shunts S and S_1 are used so as to make the two deflections equal, then,

$$\frac{c}{c_1} = \frac{\frac{G+S}{S}}{\frac{G+S_1}{S_1}}.$$

It is convenient to use Sabine's key in these measurements, taking care to adopt a uniform time of charging and a certain interval before discharging.

For capacities varying between $\frac{1}{3}$ and 1 microfarad, Dr. Muirhead recommends to charge for 15 seconds, and to allow an interval of $\frac{1}{4}$

seconds between the charge and discharge. A cable of 1,000 knots requires a charge of 5 minutes and an interval of 10 minutes.

MEASUREMENT OF ENERGY.

Energy is measured in different ways, according to the nature of the phenomena to which it gives rise and the forms under which it manifests itself. There are many classes of instruments, which are called:

Calorimeters, when the energy appears under the form of heat.
Dynamometers, when it appears under the form of mechanical work.
Ergmeters, when it appears under the form of an electric current.

The electrician has very seldom to use calorimetric methods, but it is as well to describe one method here for observing the heat produced in a wire through which a current is passing. A vessel is taken which contains a known weight of oil (in grammes), it is carefully closed and enveloped in several thicknesses of flannel or felt, to prevent loss of heat by radiation, the wire is placed in the vessel, and being immersed in oil it is insulated; the temperature t_1 of the oil is observed, the vessel is closed and placed in its envelopes and the current allowed to pass for a time T; the vessel is then opened, and the temperature of the oil t_2 rapidly observed; then if s be the specific heat of the oil, the total quantity of heat H produced in the time T is

$$H = (t_2 - t_1)s \text{ calories (g.-d.)},$$

if t_1 and t_2 be observed in centigrade degrees.

DYNAMOMETERS.

Classification.—Under the name of *dynamometers* are included all apparatus which measure the work produced or absorbed by a machine; hence there are two distinct classes: (1) *Absorption dynamometers* or *dynamometer breaks*, which measure the work produced; (2) *transmission dynamometers*, which are interposed between the motor and the machine which it drives, and which measure the work expended. These instruments are interesting to the electrician on account of their importance in testing dynamo machines and electromotors. We will describe the forms of apparatus most generally used.

Absorption dynamometers.—The simplest and best known and most used is the *Prony break*. It has been improved by *Appoldt*, *Kretz*, *Easton* and *Anderson*, *Amos*, *Emery*, *Brauer*, *Marcel Deprez*, *J. Carpentier*, *N. Raffard*, *Bramwell*, etc., who have made the apparatus easier to handle, and enabled us to obtain a certain proportionality between the coefficient of friction and the resistance, *i.e.* an automatic regulation of the instrument. There are several simple forms: A cord may be passed over a pulley and attached to the ground at one end by a spring balance, the other end carrying a scale pan, the reading on the balance minus the weight in the pan multiplied by the radius of the pulley, its circumference and the speed, gives the energy. Let R be the radius in feet, W the reading on the balance, W' the weight in the scale pan in pounds, and S the speed in revolutions per minute, then

$$\text{Activity} = (W - W') 2\pi rS \text{ foot-pounds per minute.}$$

Ayrton and *Perry* use two scale pans, the cord passing over the pulley being partly of thin smooth cord and partly of thick rough cord spliced together, the heavier weight being suspended from the thin cord, so that if the heavier weight tends to rise, the thinner cord comes on to the pulley, and thus diminishes the friction, and thus prevents the weight from being thrown over the pulley. The varying quantities of thick and thin cord form a sensitive self-adjustment of considerable range.

Transmission dynamometers.— There are a great number of these based on different principles. (1) The difference of tension of the two parts of the belt driving the machine is measured, and its speed of rotation. From this the work is deduced, after corrections for friction, slipping, elongation of the strap, etc. This class includes the dynamometers of *Froude, Parsons, Tatham, Farcot*, etc.

(2) The difference of rigidity of the two parts of the strap is measured, and from it the difference of tension is deduced. The typical instrument of this class is that of *Hefner-Alteneck*, and the modifications of it introduced by *Briggs, Elihu Thomson*, and *Hopkinson*.

(3) The motive effort is transmitted to the machine directly by means of a spring, and the value of the effort is measured, which, multiplied by the speed, gives the work. These instruments are sometimes supplied with a counter which registers the sum of the work produced during a given time. The instruments *of the Agricultural Society of London, Mégy, J. Morin, Bourry, Taurines*, etc.

The tension of the springs is sometimes measured by an optical method, as in the *Latchinoff's* dynamometer, sometimes by an index, as in *Ayrton* and *Perry's* instrument. Others act through a weight like the instruments of *Darwin, Raffard, the German dynamometer, King's, Whyte's*, etc.

(4) The work is measured by the tension of the moving axle, as in *Hirn's pandynamometer* and *Carlo Resio's* apparatus.

REVOLUTION COUNTERS AND SPEED INDICATORS.

In all dynamometrical measurements it is necessary to know the speed of rotation of the machines. This speed of rotation is measured by means of two classes of instruments which enable us to know the number of revolutions per minute, most commonly represented by the symbol n.

(1) **Revolution counters** show the mean speed of the machine during the time the experiment lasts, generally half a minute. In France *Sainte's* and *Deschien's* counters are used. When the speed does not exceed 80 to 100 revolutions per minute it is easy to count them directly without an instrument if a visible mark is made on some point of the revolving apparatus.

(2) **Speed indicators** show the speed of a machine at each instant, and this enables its regularity to be judged of. They are fixed on the axle itself or to a special transmitter. The most used are *Buss'* tachymeters, and *Jacquemier's* indicator, based on centrifugal force. *Marcel Deprez* has also constructed one based on electro-magnetic actions.

MEASUREMENT OF ELECTRICAL ENERGY.

The measurement of the energy consumed or produced by an electrical apparatus is generally made by an indirect method which consists in the measurement of two elements which concur in the production of this energy, and introducing them into a formula which gives the result sought for. There are, however, some instruments which give this result directly. We will rapidly scan the direct and indirect methods which are most used.

Energy expended by an electrical apparatus.—If C be the strength of the current which passes through the apparatus, and E the difference of potential between the terminals, the work absorbed W then is,

$$W = \frac{CE}{9\cdot 81} \text{ kilogrammes per second} = \frac{CE}{1\cdot 356} \text{ foot-pounds per second.}$$

This formula enables us to calculate the energy absorbed by an electric lamp, a motor, a resistance, etc. It springs directly from Ohm's and Joule's laws. Expressing W in horse-power,

$$W = \frac{CE}{746} \text{ horse-power, or } W = \frac{CE}{736} \text{ chevaux-vapeur.}$$

Heat produced in a conductor through which a current passes.—If R is the resistance of the conductor (at the temperature of the experiment), E the difference of potential at the extremities of this resistance, C the strength of the current passing through it, the energy W, produced in the conductor in the form of heat, is calculated by one of the following formulæ:

$$W = \frac{CE}{9 \cdot 81} = \frac{C^2 R}{9 \cdot 81} = \frac{E^2}{9 \cdot 81\, R} \text{ kilogrammes per second;}$$

$$= \frac{CE}{1 \cdot 356} = \frac{C^2 R}{1 \cdot 356} = \frac{E^2}{1 \cdot 356\, R} \text{ foot-pounds per second;}$$

or in calories (g.-d.) by the formulæ:

$$W = \frac{CE}{4 \cdot 16} = \frac{C^2 R}{4 \cdot 16} = \frac{E^2}{4 \cdot 16\, R} \text{ calories (g.-d.) per second;}$$

or $W = \cdot 2405\, CE = \cdot 2405\, C^2 R = \dfrac{\cdot 2405\, E^2}{R}$ calories (g.-d.) per second.

Ayrton and Perry's ergmeter.—A movable light coil of fine wire with small moment of inertia free to move round an axis parallel to its length, is suspended by a bifilar suspension in a fixed coil of thick wire. The fine wire is arranged as a shunt, and the thick wire in the main circuit. The deflection is a measure of the product. *Marcel Deprez* published a few years ago an analogous ergmeter, in which the action of the currents was balanced by a weight, but the indications of this instrument can only be correct if there be no relative displacement of the two circuits. It is easier in general to measure E and C separately and take their product, so that electrical ergmeters have not yet come into practical use. Recently *Vernon-Boys* has invented integrating ergmeters. These instruments show the sum of the energy absorbed by an electrical apparatus during a given time, they add up the number of kilogrammètres expended; but as yet they are rather complicated, which prevents their immediate application, therefore we only mention them.

Ayrton and *Perry* have devised a simple form of recording ergmeter intended to show the quantity of power used by a consumer from a public system of electrical supply. It consists of a fairly good clock; the pendulum bob is replaced by a flat coil of fine wire, so connected that it can be arranged as a shunt between the supply poles. Close to this coil, but fixed to the clock case, is another flat coil of very stout wire, included in the main circuit. According to the relative directions of the currents in these coils the rate of the clock is accelerated or retarded when the current is passing. This acceleration or retardation is proportional to C E, and therefore to the energy which has passed in any time. By arranging the instrument so that the loss or gain due to the passing of the currents is very much larger than the mean rate of the clock, the instrument may be made sufficiently accurate for practical purposes. Say the clock is found at the end of a month to have gained or lost five hours, a table will at once give the number of volt-ampères, or ergs, or horse-powers per hour, or foot-pounds, or kilogrammètres, which have passed through the instrument during the month.

CABLE MEASUREMENTS.

The measurement of submarine cables forms one of the most important branches of the applications of electricity; the methods employed are for the most part special. For this reason we thought it better to separate them from the general methods, and form a separate chapter for them. We will only indicate the most important methods, leaving out the question of localising faults, which would require too much detail, and for which the reader ought to consult special works. He will find a list of suitable books in the bibliography at the end of the volume.

Special arrangements.—The difficulties in carrying out the measurement of submarine cables, and the necessary precision, make it necessary to perform these measurements under special conditions which are not found in other branches of applied electricity. We will point out here the most important arrangements.

Standard temperature.—On account of the influence of temperature on the properties of the dielectric substances which are used in the construction of submarine cables, all results are reduced by calculation to a certain standard of temperature; by usage 75° Fahrenheit is adopted as the standard, which corresponds with 24° Centigrade.

Tank.—For measuring capacity and insulation the cable is placed in a tank of cast or sheet iron perfectly connected to the earth either by gas and water-pipes, or on board ship through the hull. In the case of submerged cables the cable communicates with the earth by its external metallic envelope. The cable ought to be discharged before the tests are made by connecting it to earth for some hours. It is better to bring the ends of the cable into the testing room than to use auxiliary wires.

To insulate the end of a cable.—The core is uncovered for about 40 to 50 centimètres, it is cleared of the hemp and iron wires. If the core is of indiarubber the felt is removed, and the conductor is laid bare for a length of about 3 centimètres. The conductor and the dielectric are then covered with paraffin for a total length of about 5 centimètres, and the end is kept suspended in the air.

Instruments.—Complete measurements require a battery of 500 elements, a reflecting galvanometer and its shunts, a condenser, reversing keys, short circuit keys, charging and discharging keys, commutators, a Wheatstone's bridge, and resistance boxes.

RESISTANCE OF THE CONDUCTOR.

Bridge method.—Wheatstone's bridge is generally used, especially when both ends of the cable of which the resistance is to be measured can be got at. It is then measured like an ordinary resistance. Fig. 42 shows the arrangement of the instrument.

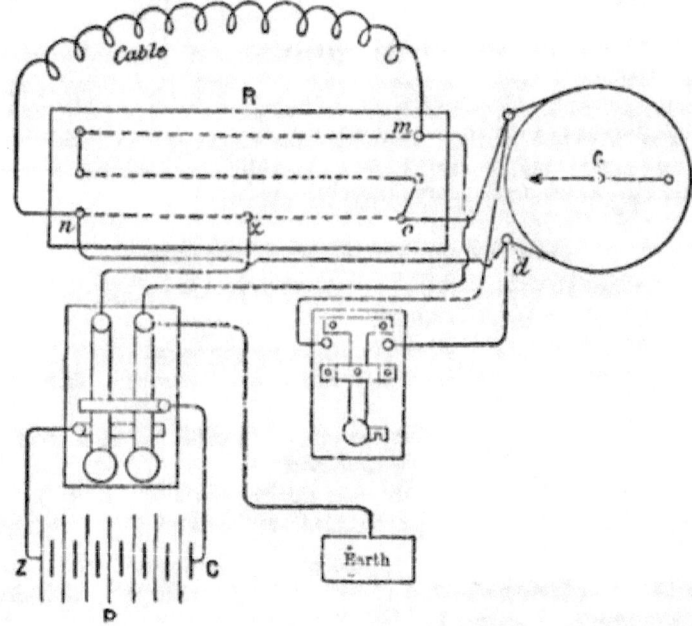

Fig. 42.—Diagram of Connections for Bridge Test of Conductor of a Cable.

False zero method.—The connections are arranged as if for a measurement by Wheatstone's bridge. The short circuit key of the galvanometer is pressed; the galvanometer deflects under the action of the earth current, and the deflection is observed. The **reversing key** is then pressed down, and the resistance box is rapidly unplugged until the same deflection is reproduced; the resistance is then read. This method can only be used when the earth current is constant.

Reproduced deflection method (*Frank Jacob*).—The cable, of which the far end is to earth, is joined to a galvanometer suitably shunted, a battery, a reversing key, and to earth; a series of readings is taken as rapidly as possible, reversing the current each time. Then a resistance box is substituted for the cable, and varied until the same deflections are reproduced. The resistances thus obtained are equal to the apparent resistances of the cable. The harmonic mean of the results of positive and negative currents gives the true resistance of the conductor. This method is very quick, and enables the variations caused by the earth currents to be easily eliminated. A battery of from 4 to 10 elements is used.

Resistance of earth plates.—A galvanometer, one battery element, a large resistance, and the earth plates are arranged in circuit, and the deflection is read. The wires are then joined directly together without the plates. The deflection ought not to change if the earths are good; if the earths have too high a resistance it may be remedied by watering the earth round the plates.

ELECTROSTATIC OR INDUCTIVE CAPACITY

is measured by the *charge* which a cable receives with **unit potential the volt**; it is expressed in *microfarads*.

The charge of a cable or condenser is proportional to the potential and the length of the cable, and inversely to the distance between the inducing surfaces.

The rate of discharge or time necessary for a cable to lose a given part of its charge is independent of the potential.

With guttapercha at 24° C. the loss during the first minute is 7 per cent.; in the second minute 7 per cent. of the remaining charge, and so on to infinity.

Loss of charge.—Let C be the initial charge, c the charge after t minutes, the charge C′ remaining after t' minutes is given by the equation

$$\frac{\log \frac{C}{C'}}{\log \frac{C}{c}} t = t'.$$

Loss of half charge.—The time in minutes is,

$$t' = t \frac{\cdot 30103}{\log \frac{C}{c}}.$$

Measurement of the electrostatic capacity of a cable.—One of the ends of the cable is insulated. The condenser is charged by a battery of 10 elements; it is discharged through the galvanometer, and the deflection noted; the cable is then charged, and discharged after one minute, shunting the galvanometer so as to obtain the same deflection; the capacity of the cable is equal to the multiplying power of the shunt multiplied by the capacity of the condenser taken as a standard. To get the largest possible deflection, and to reduce the influence of errors of reading, the right-hand extremity of the scale may be taken as zero, and the spot of light be made to deflect up to the left-hand end by varying the number of elements, and also varying the sensibility of the galvanometer by means of the directing magnet.

Capacity per kilomètre, mile, or knot is obtained by dividing the total capacity by the length of the cable in kilomètres, miles, or knots.

Calculation of the electrostatic capacity.—If D be the diameter of the dielectric in millimètres, d the diameter of the conductor, the capacity per knot or nautical mile (1852 mètres) is:

$$\text{Hooper's indiarubber:} \quad \frac{\cdot 1535}{\log \frac{D}{d}} \text{ microfarads.}$$

$$\text{Common guttapercha:} \quad \frac{\cdot 18769}{\log \frac{D}{d}} \text{ microfarads.}$$

$$\text{W. Smith's guttapercha:} \quad \frac{\cdot 15163}{\log \frac{D}{d}} \text{ microfarads.}$$

Example.—If $d = 3 \cdot 68$ mm. and $D = 8 \cdot 08$ m., the capacity per knot will be:

Hooper's indiarubber	$= \cdot 45$ microfarad.
Common guttapercha	$= \cdot 55$,,
W. Smith's guttapercha	$= \cdot 44$,,

United potential of two condensers or two cables.

If C be the capacity of the charged condenser;
V the potential of the charge;
c the capacity of the uncharged condenser;
v the potential when they are joined together;

$$v = V \frac{C}{C+c}.$$

Inductive capacity of two condensers joined together.
When a charged condenser is joined to another one which is not charged, the charge is divided between them proportionally to the respective capacities, the potential is the same in both.

If C is the standard condenser charged to the potential V, x the capacity of the second condenser, v their common potential when they are joined, then,

$$\frac{v}{V} = \frac{C}{C+x} \text{ and } CV = Cv + xv;$$

the capacity x is:

$$x = \frac{V-v}{v} C.$$

INSULATION.

Deflection method (*Latimer Clark*).—If x be the insulation of the cable, G the resistance of the galvanometer, n the number of elements of resistance r, and e. m. f. E, and a the observed deflection. The cable is removed, and replaced by a known resistance, such that the total resistance of the circuit becomes R. When the number of elements has been reduced to one and the galvanometer has been shunted by a resistance S, the deflection observed is β. Then,

$$x = R \frac{a}{\beta} n \left(1 + \frac{G}{S}\right) - (G + r) \text{ ohms.}$$

$(G + r)$ may be neglected, as x is very large; the formula then becomes,

$$x = R \frac{a}{\beta} n \left(1 + \frac{G}{S}\right) \text{ ohms.}$$

In practice the galvanometer is shunted to $\frac{1}{100}$ by a shunt $\left(S = \frac{G}{99}\right)$ and $R = 10{,}000 - (r + S)$.

The insulation resistance of the whole cable is then

$$x = \frac{a}{\beta} n \text{ megohms};$$

and its resistance per nautical mile or knot,

$$x_n = \frac{a}{\beta} \cdot \frac{1}{1852} n \text{ megohms}.$$

Differential galvanometer method (*Siemens*). — (*See* Fig. 43.)

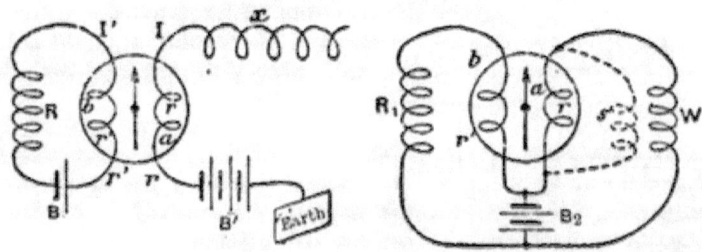

Fig. 43.—Siemens' Insulation Test

x is the insulation resistance sought;
E the e. m. f. of the battery B;
E' the e. m. f. of the battery B';
r the resistance of the battery B, together with the coil a;
r' the resistance of the battery B', together with the coil b;
R a resistance interposed in the circuit of b;

R is varied until the needle comes to zero. The cable is removed, and a known resistance W is substituted for it. The ends of the two coils a and b are joined to the pole of a single element B_2, and the resistance R_1 re-adjusted until the needle comes back to zero. The insulation of the cable is then,

$$x = \frac{W + r}{R' + r'} \cdot \frac{E}{E'} (R + r').$$

If in the second part of the experiment the coil a be shunted with a resistance s, the value of x then becomes,

$$x = \frac{W(r + s) + rs}{(R' + r')s} \cdot \frac{E}{E'} (R + r').$$

Method by loss of charge (*Siemens*).—Let C be the momentary discharge (total potential) of a given cable, c its discharge (reduced potential) after t minutes, F its capacity in microfarads. Its insulation resistance R_i after t minutes is then,

$$R_i = 26\cdot06 \frac{t}{F.\,(\log C - \log c)} \text{ megohms.}$$

Insulation of joints.—The joint to be tested is placed in a perfectly insulated tank filled with salt water, in which a plate of copper is placed. The core of the cable is carefully dried on each side of the joint. After having verified the insulation of the tank, the test is made with a battery of 500 Daniell's elements. The insulation ought not to be less than that of about 2 mètres of cable absolutely identical with that in which the joint has been made.

Calculation of insulation.—This may be calculated either by diameters or by weights. The annexed table gives the formula for the temperature of 24° C. for the dielectrics commonly used. D and d have the same significance as before. W is the weight of the dielectric, w the weight of the conductor (*V. Hoskiær*).

Example.—If $d = 2\cdot87$ mm. and $D = 7\cdot39$ mm., the insulation per knot (1,852 mètres) is:

Hooper's indiarubber	= 5340	megohms.
Common guttapercha	= 316	,,
W. Smith's guttapercha	= 144	,,

The weights W and w are expressed in kilogrammes.

Nature of the Dielectric.	CALCULATED INSULATION.	
	By Diameters.	By Weights.
	megohms.	megohms.
Hooper's indiarubber . .	$1\cdot3\ \log \frac{D}{d}$.	$1\cdot3\ \log \sqrt{1 + 5\cdot7 \frac{W}{w}}$.
Common guttapercha . .	$\cdot077 \log \frac{D}{d}$.	$\cdot077 \log \sqrt{1 + 6\cdot9 \frac{W}{w}}$.
W. Smith's guttapercha .	$\cdot035 \log \frac{D}{d}$.	$\cdot035 \log \sqrt{1 + 6\cdot9 \frac{W}{w}}$.

Speed of signalling through cables is proportional to

$$\frac{S}{l^2 C} \quad \text{or} \quad \frac{1}{l^2 Cr}.$$

S is the specific conductivity of the copper; C the capacity per knot; l length of the cable; r the resistance of the conductor. The absolute speed in number of words per minute is, with a reflecting galvanometer,

Hooper's indiarubber cable . . 1193·5 $\frac{d^2}{l^2}$ (log D − log d).

W. Smith's guttapercha . . 968·75 $\frac{d^2}{l^2}$ (log D − log d).

Common guttapercha . . . 903·65 $\frac{d^2}{l^2}$ (log D − log d).

The maximum speed obtained is 50 per cent. higher than the above figures.

Time of transmission of a signal (*R. Sabine*).—The time in seconds t for a signal to be produced at the extremity of the cable is,

With the Morse instrument . . . $t = \frac{414}{10^9} Cr$ seconds.

" Hughes " . . . $t = \frac{105}{10^9} Cr$ "

" reflecting " . . . $t = \frac{47}{10^9} Cr$ "

C total capacity of the cable in microfarads; r resistance of the conductor in ohms. The speed depends both on the inertia of the instrument and the retardation produced by the cable.

Weight of the conductor and of the dielectric.— Expressing diameters D and d in millimètres, the weights in kilogrammes will be given by the formulæ:

Conductor.	Weight per knot.	Weight per kilomètre.
Solid copper	12·78 d^2	6·889 d^2.
Stranded copper . . .	10 d^2	5·399 d^2.
Dielectric.		
Hooper's indiarubber . . .	1·75 (D^2 − d^2)	·945 (D^2 − d^2).
Guttapercha	1·43 (D^2 − d^2)	·771 (D^2 − d^2).

J

Fourth Part.

PRACTICAL INFORMATION.—APPLICATIONS.—EXPERIMENTAL RESULTS.

ALGEBRAIC FORMULÆ.

Permutations and combinations. — The different groups which can be made of n things each out of m things are called the combinations of n out of m things. Amongst these combinations are (1) the *different products* in which each group differs from the others only by one term without reckoning their order; (2) *the different permutations* in which two groups may be considered distinct only by the order of the n things which compose them. The number of permutations of m letters n by n is equal to

$$(m - n + 1)(m - n + 2) \ldots (m - 1) m.$$

The number of distinct products of m things n by n is equal to

$$\frac{1.2.3 \ldots (m-1) m}{1.2.3 \ldots n \cdot 1.2.3 \ldots (m-n)}.$$

The number of *different permutations* of m letters is equal to

$$1.2.3 \ldots (m-1) m.$$

Newton's binomial theorem. — This is so often used that we reproduce it in its most general form:

$$(x + a)^n = x^n + n a x^{n-1} + a^2 \frac{n(n-1)}{1 \cdot 2} x^{n-2} + \ldots$$
$$+ a^p \frac{n(n-1) \ldots (n-p+1)}{1.2.3 \ldots p} x^{n-p} + \ldots + a^n.$$

TABLE

Of Numbers (n); Their Reciprocals $\left(\dfrac{1}{n}\right)$; Squares ($n^2$); Square Roots $\left(\sqrt{n}\right)$

Cubes (n^3); Cube Root $\left(\sqrt[3]{n}\right)$;

Circumferences (πn); and Areas of Circles $\left(\dfrac{\pi n^2}{4}\right)$.

Where n is the Diameter.

n	$\dfrac{1}{n}$	n^2	$\sqrt{n}$	n^3	$\sqrt[3]{n}$	πn	$\dfrac{\pi n^2}{4}$
1	1	1	1·	1	1·	3·14	·79
2	·5	4	1·414	8	1·259	6·28	3·14
3	·3333	9	1·732	27	1·442	9·42	7·07
4	·25	16	2·	64	1·587	12·57	12·57
5	·2	25	2·236	125	1·709	15·71	19·63
6	·1667	36	2·449	216	1·817	18·85	28·27
7	·1429	49	2·635	343	1·912	21·99	38·48
8	·1250	64	2·828	512	2·	25·13	50·27
9	·1111	81	3·	729	2·08	28·27	63·62
10	·1	100	3·162	1,000	2·154	31·42	78·54
11	·0909	121	3·316	1,331	2·223	34·56	95·03
12	·0833	144	3·464	1,728	2·289	37·7	113·1
13	·0769	169	3·605	2,179	2·351	40·84	132·73
14	·0714	196	3·741	2,744	2·41	43·98	153·94
15	·0667	225	3·872	3,375	2·466	47·12	176·71
16	·0625	256	4·	4,096	2·519	50·27	201·06
17	·0588	289	4·123	4,913	2·571	53·41	226·98
18	·0556	324	4·242	5,832	2·62	56·55	254·47
19	·0526	361	4·358	6,859	2·668	59·69	283·53
20	·05	400	4·472	8,000	2·714	62·83	314·16
21	·0476	441	4·582	9,261	2·758	65·97	346·36
22	·0455	484	4·69	10,648	2·802	69·11	380·13
23	·0435	529	4·795	12,167	2·843	72·26	415·48
24	·0417	576	4·898	13,824	2·884	75·4	452·39
25	·04	625	5·	15,625	2·924	78·54	490·87
26	·0385	676	5·099	17,576	2·962	81·68	530·93
27	·037	729	5·196	19,683	3·	84·82	572·56
28	·0357	784	5·291	21,952	3·036	87·96	615·75
29	·0345	841	5·381	24,389	3·072	91·11	660·52
30	·0333	900	5·477	27,000	3·107	94·25	706·86
31	·0323	961	5·567	29,791	3·141	97·39	754·77
32	·0313	1,024	5·656	32,768	3·174	100·53	804·25
33	·0303	1,089	5·744	35,937	3·207	103·67	855·30
34	·0294	1,156	5·830	39,304	3·239	106·81	907·92
35	·0286	1,225	5·916	42,875	3·271	109·96	962·11

n	$\dfrac{1}{n}$	n^2	$\sqrt{n}$	n^3	$\sqrt[3]{n}$	πn	$\dfrac{\pi n^2}{4}$
36	·0278	1,296	6·	46,656	3·391	113·1	1,017·88
37	·0270	1,369	6·082	50,653	3·332	116·24	1,075·21
38	·0263	1,444	6·164	54,872	3·361	119·38	1,134·11
39	·0256	1,521	6·244	59,319	3·391	122·52	1,194·59
40	·025	1,600	6·324	64,000	3·419	125·66	1,256·64
41	·0244	1,681	6·403	68,921	3·448	128·8	1,320·25
42	·0238	1,764	6·48	74,088	3·476	131·95	1,385·44
43	·0233	1,849	6·557	79,507	3·503	135·09	1,452·2
44	·0227	1,936	6·633	85,184	3·530	138·23	1,520·53
45	·0222	2,025	6·708	91,125	3·556	141·37	1,520·43
46	·0217	2,116	6·782	97,336	3·583	144·51	1,661·9
47	·0213	2,209	6·855	103,823	3·608	147·65	1,734·94
48	·0208	2,304	6·928	110,592	3·634	150·8	1,809·56
49	·0204	2,401	7·	117,649	3·659	153·94	1,885·74
50	·02	2,500	7·071	125,000	3·684	157·08	1,963·49
51	·0196	2,601	7·141	132,651	3·708	160·22	2,042·82
52	·0192	2,704	7·211	140,608	3·732	163·36	2,123·72
53	·0189	2,809	7·28	148,877	3·756	166·5	2,206·18
54	·0185	2,916	7·348	157,464	3·779	169·65	2,290·21
55	·0182	3,025	7·416	166,375	3·802	172·79	2,375·83
56	·0179	3,136	7·483	175,616	3·825	175·93	2,463·01
57	·0175	3,249	7·549	185,193	3·848	179·07	2,551·76
58	·0172	3,364	7·615	195,112	3·87	182·21	2,642·08
59	·0169	3,481	7·681	205,379	3·892	185·35	2,733·97
60	·0167	3,600	7·745	216,000	3·914	188·5	2,827·43
61	·0164	3,721	7·81	226,981	3·936	191·64	2,922·47
62	·0161	3,844	7·874	238,328	3·957	194·78	3,019·07
63	·0159	3,969	7·937	250,047	3·979	197·92	3,117·24
64	·0156	4,096	8·	262,144	4·	201·06	3,216·99
65	·0154	4,225	8·062	274,625	4·02	204·2	3,318·31
66	·0152	4,356	8·124	287,496	4·041	207·34	3,421·19
67	·0149	4,489	8·185	300,763	4·061	210·49	3,525·65
68	·0147	4,624	8·246	314,432	4·081	213·63	3,631·68
69	·0145	4,761	8·306	328,509	4·101	216·77	3,739·28
70	·0143	4,900	8·366	343,000	4·121	219·91	3,848·45
71	·0141	5,041	8·426	357,911	4·14	223·05	3,959·19
72	·0139	5,184	8·485	373,248	4·16	226·19	4 071·5
73	·0137	5,329	8·544	389,017	4·179	229·34	4,185·39
74	·0135	5,476	8·602	405,224	4·198	232·48	4,300·84
75	·0133	5,625	8·66	421,875	4·217	235·62	4,417·86
76	·0132	5,776	8·717	438,976	4·235	238·76	4,536·46
77	·013	5,929	8·774	456,533	4·254	241·9	4,656·62
78	·0128	6,084	8·831	474,552	4·272	245·04	4,778·36
79	·0127	6,241	8·888	493,039	4·29	248·19	4,901·67
80	·0125	6,400	8·944	512,000	4·308	251·33	5,026·55
81	·0123	6,561	9·	531,441	4·326	254·47	5,153·
82	·0122	6,724	9·055	551,368	4·344	257·61	5,281·02
83	·012	6,889	9·11	571,787	4·362	260·75	5,410·61

n	$\dfrac{1}{n}$	n^2	$\sqrt{n}$	n^3	$\sqrt[3]{n}$	πn	$\dfrac{\pi n^2}{4}$
84	·0119	7,056	9·165	592,704	4·379	263·89	5,541·77
85	·0118	7,225	9·219	614,125	4·396	267·03	5,674·5
86	·0116	7,396	9·273	636,056	4·414	270·18	5,808·8
87	·0115	7,569	9·327	656,503	4·431	273·32	5,944·68
88	·0114	7,744	9·386	681,472	4·447	276·46	6,082·12
89	·0112	7,921	9·433	704,969	4·464	279·6	6,221·14
90	·0111	8,100	9·486	729,000	4·481	282·74	6,361·72
91	·011	8,281	9·539	753,571	4·497	285·88	6,503·88
92	·0109	8,464	9·591	778,688	4·514	289·03	6,647·61
93	·0108	8,649	9·643	804,357	4·530	292·17	6,792·91
94	·0106	8,836	9·695	830,584	4·546	295·31	6,939·78
95	·0105	9,025	9·746	857,375	4·562	298·45	7,088·22
96	·0104	9,216	9·797	884,736	4·578	301·59	7,238·23
97	·0103	9,409	9·848	912,673	4·594	304·73	7,389·81
98	·0102	9,604	9·899	941,192	4·61	307·88	7,542·96
99	·0101	9,801	9·949	970,229	4·626	311·02	7,697·69
100	·01	10,000	10·	1,000,000	4·642	314·16	7,853·98

π *and functions of* π.

$\pi = 3{\cdot}1415926536.$

$\dfrac{1}{\pi} = 0{\cdot}3183098862.$

$\pi^2 = 9{\cdot}8696.$

$\pi^3 = 31{\cdot}0063.$

$\sqrt{\dfrac{1}{\pi}} = 0{\cdot}56419.$

$\log \pi = 1{\cdot}4971498727.$

$\log \dfrac{1}{\pi} = 1{\cdot}5028501273.$

$\sqrt{\pi} = 1{\cdot}77245385.$

$\sqrt[3]{\pi} = 1{\cdot}4646.$

$\log_e \pi = 1{\cdot}14473.$

Diameter of circle of which the circumference is one
mètre . . . . = 31·83 cm.
Length of arc of 1° and radius 1 , ... = ·017452.
Radian angle of which the arc is equal to the radius
(the unit of circular measure) . . . . = 57° 15′.

Arithmetical progression.—Let a be the first term; r the difference between any two terms; b the last term; and n the number of terms.

$$b = a + (n-1)r.$$

Sum of first n terms $= \dfrac{[2a + (n-1)r]n}{2} = \dfrac{a+b}{2}n.$

Sum of first n numbers from 1 to $n = \dfrac{n(n+1)}{2}.$

Sum of first n odd number from 1 to $(2n-1) = n^2.$

Geometrical progression.—Let a be the first term, b the last term, and q the ratio of any term to the next. Then,

$$b = aq^{n-1}.$$

Sum of the first n terms $= a\,\dfrac{q^n - a}{q - 1}.$

Logarithms.

If there be three numbers a, b, x, such that

$$a^x = b,$$

is called the logarithm of the number b to the base a, and we write:

$$x = \log b.$$

The number b is the *antilogarithm*.

The collection of logarithms of the different numbers to one and the same base form a *system of logarithms*. Only two systems are now used in practice.

1st. *Common* or *decimal logarithms*, of which the base is 10.

2nd. *Natural, Naperian*, or *hyperbolic logarithms*, of which the base is e.

$$e = 2\cdot 71828.$$

Common logarithms are indicated by the symbol (log); natural logarithms by ($\log_e$).

TABLE

Of Decimal (log) and Naperian (log$_e$) Logarithms of Numbers from 1 to 100.

Numbers.	Logarithms. Decimal.	Logarithms. Naperian.	Numbers.	Logarithms. Decimal.	Logarithms. Naperian.
1	0·	0·	36	1·55630	3·58351
2	·30103	·69314	37	1·56820	3·61091
3	·47712	1·09861	38	1·57978	3·63758
4	·60206	1·38629	39	1·59106	3·66356
5	·69897	1·60943	40	1·60206	3·68887
6	·7,815	1·79175			
7	·84510	1·94591	41	1·61278	3·71357
8	·90309	2·07944	42	1·62325	3·73766
9	·95424	2·19722	43	1·63347	3·76120
10	1·00000	2·30258	44	1·64345	3·78418
			45	1·65321	3·80666
11	1·04139	2·39789	46	1·66276	3·82864
12	1·07918	2·48490	47	1·67210	3·85014
13	1·11394	2·56494	48	1·68124	3·87120
14	1·14613	2·63905	49	1·69020	3·89182
15	1·17609	2·70805	50	1·69897	3·91202
16	1·20412	2·77258			
17	1·23045	2·83321	51	1·70757	3·93182
18	1·25527	2·89037	52	1·71600	3·95124
19	1·27875	2·94413	53	1·72428	3·97029
20	1·30103	2·99573	54	1·73239	3·98898
			55	1·74036	4·00733
21	1·32222	3·04452	56	1·74819	4·02535
22	1·34242	3·09104	57	1·75587	4·04305
23	1·36173	3·13549	58	1·76343	4·06044
24	1·38021	3·17805	59	1·77085	4·07753
25	1·39794	3·21887	60	1·77815	4·09434
26	1·41497	3·25809			
27	1·43136	3·29583	61	1·78533	4·11087
28	1·44716	3·33220	62	1·79239	4·12713
29	1·46240	3·36729	63	1·79934	4·14313
30	1·47712	3·40119	64	1·80618	4·15888
			65	1·81291	4·17439
31	1·49136	3·43398	66	1·81954	4·18965
32	1·50515	3·46573	67	1·82607	4·20469
33	1·51851	3·49650	68	1·83251	4·21951
34	1·53148	3·52636	69	1·83885	4·23411
35	1·54407	3·55534	70	1·84510	4·24850

NUMBERS.	LOGARITHMS.		NUMBERS.	LOGARITHMS.	
	DECIMAL.	NAPERIAN.		DECIMAL.	NAPERIAN.
71	1·85126	4·26268	86	1·93450	4·45435
72	1·85733	4·27667	87	1·93952	4·46591
73	1·86332	4·29046	88	1·94448	4·47734
74	1·86923	4·30407	89	1·94939	4·48864
75	1·87506	4·31749	90	1·95424	4·49981
76	1·88081	4·33073			
77	1·88649	4·34381	91	1·95904	4·51086
78	1·89209	4·35671	92	1·96379	4·52170
79	1·89763	4·36945	93	1·96848	4·53260
80	1·90309	4·38203	94	1·97313	4·54329
			95	1·97772	4·55388
81	1·90849	4·39445	96	1·98227	4·55435
82	1·91381	4·40672	97	1·98677	4·56471
83	1·91908	4·41884	98	1·99123	4·58497
84	1·92428	4·43082	99	1·99564	4·59512
85	1·92942	4·44265	100	2·00000	4·60517

Knowing the log of a number n in the one system, its log in the other system is obtained by the following formula:

$$\log_e n = 2·3025851 \log n.$$
$$\log n = 0·434294482 \log_e n.$$

Properties of logs.—Those which are most generally used are set out in the following table:

$$\log ab = \log a + \log b.$$

$$\log \frac{a}{b} = \log a - \log b.$$

$$\log a^b = b \log a.$$

$$\log \sqrt[b]{a} = \frac{\log a}{b}.$$

Interest.—*Simple interest.*—Let i be the rate (sum brought in by one pound during one year expressed as a fraction of a pound), A the sum placed at interest during n years, which will bring in Ain pounds, and will become A $(1 + in)$.

Discount.—If a sum A be paid n years before it is due only A $(1 - in)$ is paid.

Present value.—A sum A due in n years is represented at the present time by a sum A′, such that if it were placed at the rate of interest i during the time n, it would become equal to A; therefore,

$$A' = \frac{A}{1+in}.$$

Compound interest.—A sum A placed at compound interest for n years becomes

$$A' = A(1+i)^n.$$

Annuities.—A sum a due annually for n years has a present value A given by the equation

$$A = \frac{a}{i}\left[1 - \frac{1}{(1+i)^n}\right].$$

GEOMETRICAL FORMULÆ.
Length.

SQUARE. Diagonal $= s\sqrt{2} = 1{\cdot}414\,s$ (s length of side of square).

CIRCUMFERENCE $= 2\pi r$ (r radius).

ARC OF $n° = \dfrac{\pi n r}{180}$.

Area.

CIRCLE $= \pi r^2 = \dfrac{\pi d^2}{2}$ ($r =$ radius, $d =$ diameter).

ELLIPSE $= \pi ab$ ($a =$ semi-major axis, $b =$ semi-minor axis).

CYLINDER (lateral area) sl ($s =$ circumference, l length of generatrix, or $2\pi rl$, $r =$ radius).

CONE (lateral area) $\dfrac{sl}{2}$ ($s =$ circumference, $l =$ length of generatrix, or $\pi r\sqrt{h^2 + r^2}$. $r =$ radius of base, $h =$ height of cone).

FRUSTRUM OF CONE (lateral area) $= \left(\dfrac{s+s'}{2}\right)l$ s and s' circumference of base and top, $l =$ generatrix.

SPHERE $= 4$ great circles $= 4\pi r^2$.

ZONE ON A SPHERE $= 2\pi rh$ ($h =$ width of the zone measured on the surface of the sphere).

Volume or cubic contents.

Pyramid and Cone $= \dfrac{Bh}{3}$ (B area of base, h height).

Cylinder or Prism $= Bh$.

Truncated Pyramid with the top plane parallel to the base $= \tfrac{1}{3} h \, (B + B' + \sqrt{BB'})$ (B and B' areas of base and top plane).

Frustrum of Cone with top plane parallel to base $= \tfrac{1}{3} \pi h \, (r^2 + r'^2 + rr')$ (r and r' radii of base and top plane).

Sphere $= \tfrac{4}{3} \pi r^3 = 4{\cdot}19 r^3$.

Segment of a Sphere. $-\left(\dfrac{B+B'}{2}\right) h + \dfrac{1}{6} \pi h^3$ (B and B' areas of the plane surfaces, h width of the zone).

Ellipsoid $= \tfrac{4}{3} \pi abc$ (a, b, c, semi-axes of ellipsoid).

The surface of revolution, generated by an area S in the plane of the axis of rotation, encloses a volume equal to the product of the area S, by the circumference traced out by its centre of gravity. If d be the distance of the centre of gravity from the axis of rotation, the volume of the solid generated V is

$$V = 2\pi ds.$$

RADII AND AREAS OF REGULAR INSCRIBED POLYGONS.

IN TERMS OF THE SIDE.

NAME OF POLYGON.	NUMBER OF SIDES.	INTERIOR ANGLE.	APOTHEME.	RADIUS.	AREA.
Equilateral triangle	3	60°	·288	·576	·433
Square	4	90	·5	·706	1
Pentagon	5	108	·688	·850	1·72
Hexagon	6	120	·866	1	2·59
Octagon	8	135	1·207	1·305	4·838
Decagon	10	144	1·539	1·618	7·694
Dodecagon	12	150	1·866	1·93	11·196

TABLE OF SINES AND TANGENTS.

Degrees.	Sines.	Tangents.	Degrees.	Sines.	Tangents.
1	·017	·017	46	·719	1·036
2	·035	·035	47	·731	1·073
3	·052	·052	48	·743	1·111
4	·07	·07	49	·755	1·151
5	·087	·088	50	·766	1·192
6	·105	·105	51	·777	1·236
7	·122	·123	52	·788	1·281
8	·139	·141	53	·799	1·328
9	·157	·159	54	·809	1·377
10	·174	·177	55	·819	1·429
11	·191	·195	56	·829	1·483
12	·208	·213	57	·839	1·541
13	·225	·231	58	·848	1·601
14	·242	·250	59	·857	1·665
15	·259	·268	60	·866	1·733
16	·276	·287	61	·875	1·805
17	·293	·306	62	·883	1·882
18	·309	·325	63	·891	1·964
19	·326	·345	64	·899	2·052
20	·342	·364	65	·906	2·146
21	·359	·384	66	·914	2·248
22	·375	·404	67	·921	2·358
23	·391	·425	68	·927	2·477
24	·407	·445	69	·934	2·607
25	·423	·467	70	·94	2·75
26	·439	·488	71	·946	2·907
27	·454	·510	72	·951	3·081
28	·47	·532	73	·956	3·274
29	·485	·555	74	·961	3·481
30	·5	·578	75	·966	3·736
31	·515	·601	76	·97	4·016
32	·53	·625	77	·974	4·337
33	·545	·650	78	·978	4·711
34	·559	·675	79	·982	5·152
35	·574	·701	80	·985	5·681
36	·588	·727	81	·988	6·326
37	·602	·754	82	·990	7·130
38	·616	·782	83	·992	8·164
39	·629	·810	84	·994	9·541
40	·643	·839	85	·996	11·468
41	·656	·870	86	·997	14·361
42	·669	·901	87	·998	19·188
43	·682	·933	88	·999	28·877
44	·695	·966	89	·999	58·261
45	·707	1·	90	1·	∞

TRIGONOMETRICAL FORMULÆ.

$$\sin a = 2 \sin \frac{1}{2} a \cos \frac{1}{2} a;$$

$$\tan a = \frac{\sin a}{\cos a}; \qquad \cot a = \frac{1}{\tan a}$$

$$\tan a = \frac{1 - \cos 2a}{\sin 2a}$$

$$\sin^2 a + \cos^2 a = 1$$

$$\sin a = \pm \frac{\tan a}{\sqrt{1 + \tan^2 a}}$$

$$\cos a = \pm \frac{1}{\sqrt{1 + \tan^2 a}}$$

$$\sin 2a = 2 \sin a \cos a$$

$$\cos 2a = \cos^2 a - \sin^2 a$$

$$\tan 2a = \frac{2 \tan a}{1 - \tan^2 a}$$

$$\cos a = 1 - 2 \sin^2 \frac{1}{2} a$$

$$\cos a = -1 + 2 \cos^2 \frac{1}{2} a$$

$$1 + \tan^2 a = \frac{1}{\cos^2 a}$$

$$\tan^2 a = \frac{1 - \cos 2a}{1 + \cos 2a}$$

$$\sin \frac{1}{2} a = \pm \sqrt{\frac{1 - \cos a}{2}}$$

$$\cos \frac{1}{2} a = \pm \sqrt{\frac{1 + \cos a}{2}}$$

$$\tan \frac{1}{2} a = \frac{-1 \pm \sqrt{1 + \tan^2 a}}{\tan a}$$

$$\sin(a+b) = \sin a \cos b + \cos a \sin b$$
$$\sin(a-b) = \sin a \cos b - \cos a \sin b$$
$$\cos(a+b) = \cos a \cos b - \sin a \sin b$$
$$\cos(a-b) = \cos a \cos b + \sin a \sin b$$

$$\tan(a+b) = \frac{\tan a + \tan b}{1 - \tan a \tan b}$$

$$\tan(a-b) = \frac{\tan a - \tan b}{1 + \tan a \tan b}$$

$$\sin a + \sin b = 2 \sin \frac{a+b}{2} \cos \frac{a-b}{2}$$

$$\sin a - \sin b = 2 \sin \frac{a-b}{2} \cos \frac{a+b}{2}$$

$$\cos a + \cos b = 2 \cos \frac{a+b}{2} \cos \frac{a-b}{2}$$

$$\cos a - \cos b = -2 \sin \frac{a+b}{2} \sin \frac{a-b}{2}$$

$$\tan(a+b) = \frac{\tan a + \tan b}{1 - \tan a \tan b}$$

$$\tan(a-b) = \frac{\tan a - \tan b}{1 + \tan a \tan b}$$

$$\sin a \sin b = \frac{1}{2} \cos(a-b) - \frac{1}{2} \cos(a+b)$$

$$\cos a \cos b = \frac{1}{2} \cos(a-b) + \frac{1}{2} \cos(a+b)$$

$$\sin a \cos b = \frac{1}{2} \sin(a-b) + \frac{1}{2} \sin(a+b)$$

$$\frac{\sin a \pm \sin b}{\cos a + \cos b} = \tan \frac{a \pm b}{2}$$

$$\frac{\sin a \pm \sin b}{\cos a - \cos b} = -\cot \frac{a \pm b}{2}$$

For $a + b + c = 180°$, we have:

$$\sin a + \sin b + \sin c = 4 \cos \frac{1}{2} a \cos \frac{1}{2} b \cos \frac{1}{2} c$$

SOLUTION OF TRIANGLES.

A B C are the angles; $a\ b\ c$ the respective opposite sides.

Right-angled triangles.—The hypothenuse is c, and the right angle C.

$$A = 90° - B; \quad B = 90° - A.$$

First case.—Given the hypothenuse and one angle (c, and A),

$$a = c \sin A; \quad b = c \cos A.$$

Second case.—Given the hypothenuse and one side (c, a),

$$\sin A = \frac{a}{c}; \quad b = a \cot a = \sqrt{(c+a)(c-a)}.$$

Third case.—Given one side and one angle (a, B) or (a, A),

$$c = a \cos B = a \sin A;$$
$$b = a \tan B = a \cot A.$$

Fourth case.—Given two sides (a, b),

$$\tan A = \frac{a}{b}; \quad c = \frac{a}{\sin A};$$
$$c^2 = a^2 + b^2.$$

Any triangle.

$$A + B + C = 180°;$$
$$\frac{a}{\sin A} = \frac{b}{\sin B} = \frac{c}{\sin C};$$
$$c^2 = a^2 + b^2 - 2ab \cos C.$$

First case.—Given one side and two angles,

third angle = 180° − the sum of the other two.

$$a = \frac{b}{\sin B} \sin A; \quad b = \frac{a}{\sin A} \sin B;$$
$$= \frac{b}{\sin B} \sin C = \frac{a}{\sin A} \sin C.$$

SOLUTION OF TRIANGLES.

Second case.—Given two sides and the angle opposite to one of them (a, b, A).

$$\sin B = \frac{b \sin A}{a};$$

$$C = 180° - (A + B);$$

$$c = \frac{a \sin C}{\sin A}.$$

Third case.—Given two sides and the angle included between them (a, b, C).

$$\tan \tfrac{1}{2}(A - B) = \frac{a - b}{a + b} \cot \tfrac{1}{2} C.$$

Knowing $\tfrac{1}{2}(A - B)$ we then get

$$A = (90° - \tfrac{1}{2} C) + \tfrac{1}{2}(A - B);$$

$$B = (90° - \tfrac{1}{2} C) - \tfrac{1}{2}(A - B);$$

$$c = \frac{a \sin C}{\sin A} = \frac{b \sin C}{\sin B} = \frac{(a + b) \sin \tfrac{1}{2} C}{\cos \tfrac{1}{2}(A - B)}.$$

Fourth case.—Given the three sides (a, b, c).

$$\text{Let } m = \frac{a + b + c}{2};$$

then

$$\sin \tfrac{1}{2} A = \sqrt{\frac{(m - b)(m - c)}{bc}};$$

$$\cos \tfrac{1}{2} A = \sqrt{\frac{m(m - a)}{bc}};$$

$$\sin \tfrac{1}{2} B = \sqrt{\frac{(m - a)(m - c)}{ac}}.$$

Similarly for the other angles B and C.

COINS OF DIFFERENT COUNTRIES.

Belgium, Greece, Italy, Switzerland.—These four nations since 1880 have formed a union for gold and silver coins, based on the franc ·04 of £1, 1fr. = 100 centimes.

Germany.—The *Reichs-mark* of 100 *pfennigs* = 1fr. ·2345, or nearly 1s. Gold coins, 20, 10, and 5 mark pieces: silver coins, 5, 2, 1, ½, and ⅕ marks.

Austria.—*Florin* of 100 kreutzers = 2·4691fr., or nearly 2s. 8 florins = 20fr.; about 16s. 1 ducat = 11·85fr.; about 9s. 5¼d.

Spain.—1 *peseta* = 1fr.; a little less than 10d. 1 *real* = ¼ peseta.

Holland.—*Florin* of 100 cents = 2·10fr.; about 1s. 10d.

Portugal.—*Milreïs* = 5·6fr.; nearly 4s. 6d.

Russia.—*Rouble* = 100 kopecks = 4fr.; about 1s. 5½d.

Sweden and Norway.—*Krona* = 100 ore = 1·3888fr. = 10d.

India.—*Rupee* = 2·3757fr.; less than 2s.

United States.—*Dollar* of 100 cents = 5·1825fr.; about 4s.

Brazil.—*Milreïs* = 2·83fr. = 2s. 3d.

PHYSICAL FORMULÆ.

Formulæ for falling bodies and the pendulum.

t = time in seconds.
s = space passed over in time t.
v = velocity at end of time t.
h = height from which the body falls.
g = acceleration due to gravity.
l = length of pendulum.

$$v = gt; \quad v = \sqrt{2gh};$$
$$s = \tfrac{1}{2} gt^2; \quad t = \pi \sqrt{\tfrac{l}{g}}.$$

Moment of inertia.—The *moment of inertia* of a body is the sum of the products of the masses of its every material point into the square of the distance of that point from the axis of rotation. Calling it I,

$$I = \Sigma mr^2.$$

The *radius of gyration* is the distance from the axis of rotation at which the whole mass of the body, concentrated at a point, would have the same moment of inertia as the body. If M be the whole mass of the body, and R the radius of gyration,

$$MR^2 = \Sigma mr^2. \quad \text{Whence } R^2 = \frac{\Sigma mr^2}{M}.$$

In the case of homogeneous bodies, when the radius of gyration has been determined, the moment of inertia can be deduced from it.

STRAIGHT ROD.—Let l be the length of the rod. If the axis of rotation is perpendicular at one end,

$$R^2 + \frac{l^2}{3}.$$

If the axis is perpendicular and in the middle,

$$R^2 = \frac{l^2}{12}.$$

ROD BENT INTO THE ARC OF A CIRCLE, axis passing through the centre of the arc of radius r and perpendicular to its plane,

$$R^2 = r^2.$$

Axis passing through the middle of the arc along a radius,

$$R^2 = \frac{r^2}{2}$$

DISC.—Axis passing through the centre, perpendicular to its plane,

$$R^2 = \frac{r^2}{2}.$$

Axis passing through a diameter,

$$R^2 = \frac{r^2}{4}.$$

CONE.—Radius of base r; axis passing through the axis of figure,

$$R^2 = \frac{3r^2}{10}.$$

K

SPHERE of radius r, axis passing along a diameter,

$$R^2 = \frac{2}{51} r^2.$$

Formula of the bifilar suspension.—Let l be the length of the suspending threads or wires, a their distance apart at their upper ends, b their distance apart at their lower ends, W the weight which they bear, α the angle of deflection. The moment of the couple exerted by the bifilar suspension when its position of equilibrium is changed, is,

$$M = \frac{ab}{l} \times \frac{W}{g} \sin \alpha.$$

For very small angles the sine is equal to the arc, and the moment becomes proportional to the deflection. In practice the moment is varied, and thus the sensibility changed by altering a.

VELOCITIES.

Velocity of sound in mètres per second and feet per second:

	Mètres per second.	Feet per second.
In air at 0°	330·9	1085·7
„ 10°	337·2	1106·3
In water at 8°	1435	4708·2
In cast iron	3480	11417·9

The increase in the velocity of sound in air is ·626 mètre (2 feet) per degree C.

Velocity of light:

	Kilomètres per second.	Miles per second.
Foucault (1862)	298,000	185,171
Cornu (1874)	300,000	186,414

Velocity of wind:

	Mètres per second.	Feet per second.	Pressure in Kilogrammes per square inch.	In pounds per square foot.
Fresh breeze	7	23	6	1·23
Strong wind	15	49·2	30	6·15
Storm	24	78·7	78	16
Hurricane	45	147·6	273	56·4

Mean velocity of engine straps:

	Mètres per sec.	Feet per sec.
		ft. in.
Lower limit for transmission of power by pulleys and straps	1·1	3 7
Velocity for small powers	1·5	4 11
Limit of speed for transmission of energy by ropes and cords	25	83
Velocity of large straps	6·1	20
Limit of velocity for large straps	9·15	30

Velocity of translation of the armatures of some dynamo machines:

The five-light *Gramme* machine, revolving at 1300 revolutions per minute, with a ring 26 cm. (10·4 inches) in diameter, has a velocity of 17·7 mètres per second (58 feet per second).

In the new 12-pole machines, the large size of the ring gives the high velocity of 43 mètres per second (141 feet per second) for the outside of the ring.

In the *Ferranti* machine, the speed of the middle part of the armature is 38 mètres per second (124 feet per second), and of the outside part 54 mètres per second (177 feet per second).

In the *Siemens* machines, the small armatures only move over about 8 to 10 mètres per second (26 to 32 feet per second). The large alternating current machines (no iron in the armatures) have as high a speed as 32 mètres per second (105 feet per second).

Other conditions remaining the same, the higher the speed of translation the lighter the machine, and the more its internal resistance can be diminished.

Work produced by men and horses.

A man turning a handle can work 8 hours a day, and produce 6 kilogrammètres per second (43·4 foot-pounds per second). His total daily work is therefore 172,800 kilogrammètres (1,249,879 foot-pounds). A draught-horse walking and dragging a carriage or a boat at the rate of 1·1 mètres per second (3 feet 7 inches per second) produces 59·4 kgms. per second (429·6 foot-pounds per second), so that in 8 hours it produces 1,712,000 kgms. (123,830,672 foot-pounds). A horse trotting or cantering, dragging a light carriage on rails at the rate of 4·37 mètres per second (14 feet per second) can produce 61 kgms. per second (441 foot-pounds per second) for 4 hours, or a daily work of 881,280 kgms. (6,374,386 foot-pounds).

SPECIFIC GRAVITIES (*Wurtz and Rankine*).

Weight in Grammes of one Cubic Centimètre at 0° C.

Metals.

Iridium	. . . .	22·38
Platinum	. . 21 to	22
Gold	. . . 19 to	19·6
Lead	. . . .	11·4
Silver	. . . .	10·5
Bismuth	. . . .	9·82
Copper (hammered)	. .	8·9
,, (rolled)	. . .	8·8
,, (cast)	. . .	8·6
Cadmium (rolled)	. .	8·69
Nickel (cast)	. . .	8·57
Brass (cast)	. . 7·8 to	8·4
Brass (wire)	. . .	8·54
Steel	. . 7·8 to	7·9
Wrought iron	. . .	7·8
Tin	. . 7·3 to	7·5
Zinc	. . . .	7·19
Cast iron	. . . .	7·
Selenium (black)	. .	4·8
,, (red)	. .	4·5
Aluminium (rolled)	. .	2·67
Magnesium	. . .	1·74
Sodium	. . . .	·97
Lithium	. . . .	·59

Insulators.

Flint	. . . 3 to	3.5
Crown	. . . .	2·5
Green glass	. . .	2·64
Plate ,,	. . . .	2·8
Marble	. . . .	2·7
Paraffin	. . . .	·87
Quartz	. . . .	2·65
Porcelain	. . 2·15 to	2·3
Ivory	. . . .	1·8
Silica	. . . .	1·7
Pitch	. . . .	1·65
Tar	. . . .	1·02
Hooper's indiarubber	.	1·18
Guttapercha	. ·97 to	·98
Indiarubber	. . .	·93
Ebonite	. . . .	1·15
Sulphur (octohedral)	.	2·07
,, (prismatic)	.	1·97

Liquids.

Mercury	. . . .	13·596
Bromine (at 15°)	. .	2·99
Carbon bisulphide	. .	1·263
Sea-water	. . .	1·026
Water at 4°	. . .	1
Olive oil	. . . .	·915
Naphtha	. . . .	·848
Alcohol (pure)	. . .	·791
Petroleum	. . . .	·878
Æther	. . . .	·716

Various substances.

Carré's carbon	. . .	1·62
Retort carbon	. . .	1·91
Diamond	. . . .	3·5
Coke	. . . 1 to	1·66
Ice at 4°	. . . .	·92
Loose snow	. . .	·1

SPECIFIC GRAVITIES.

TABLE OF THE DEGREES OF BAUMÉ, CARTIER, AND GAY-LUSSAC.

For Liquids lighter than Water, with their corresponding Specific Gravities.

The Gay-Lussac degrees correspond to the percentage by volume in a mixture of water and alcohol at 15° C. (*Agenda du Chimiste*).

\multicolumn{3}{c}{DEGREES.}	SPECIFIC GRAVITY.	\multicolumn{3}{c}{DEGREES.}	SPECIFIC GRAVITY.
BAUMÉ.	CARTIER.	GAY-LUSSAC.		BAUMÉ.	CARTIER.	GAY-LUSSAC.	
10	10	0	1			31	·965
		1	·999		15	32	·964
		2	·997			33	·963
		3	·996	16		34	·962
		4	·994			35	·96
11	11	5	·993			36	·959
		6	·992		16	37	·957
		7	·99			38	·956
		8	·989	17		39	·954
		9	·988			40	·953
12		10	·987		17	41	·951
	12	11	·986			42	·949
		12	·984	18		43	·948
		13	·983			44	·946
		14	·982			45	·945
		15	·981		18	46	·943
		16	·98	19		47	·941
13		17	.979			48	·94
	13	18	·978			49	·938
		19	·977	20	19	50	·936
		20	·976			51	·934
		21	·975			52	·932
		22	·974	21	20	53	·93
14		23	·973			54	·928
		24	·972			55	·926
	14	25	·971	22	21	56	·924
		26	·97			57	·922
		27	·969			58	·92
		28	·968	23	22	59	·918
15		29	·967			60	·915
		30	·966			61	·913

DEGREES.			SPECIFIC GRAVITY.	DEGREES.			SPECIFIC GRAVITY.
BAUMÉ.	CARTIER.	GAY-LUSSAC.		BAUMÉ.	CARTIER.	GAY-LUSSAC.	
24	23	62	·911			82	·86
		63	·909	34	32	83	·857
25		64	·903	35		84	·854
	24	65	·904		33	85	·851
		66	·902	36	34	86	·848
26		67	·899			87	·845
	25	68	·896	37	35	88	·842
27		69	·893	38	36	89	·838
	26	70	·891			90	·835
28		71	·888	39	37	91	·832
	27	72	·886			92	·829
29		73	·884	40	38	93	·826
	28	74	·881	41		94	·822
30		75	·879	42	39	95	·818
		76	·876	43	40	96	·814
31	29	77	·874	44	41	97	·81
		78	·871	45	42	98	·805
32	30	79	·868	46	43	99	·8
		80	·865	47	44	100	·795
33	31	81	·863	48			·791

NOTE.—If the temperature be $(15 + n)°$ (·4n) degrees must be subtracted in order to obtain the percentage of alcohol. (·4n) must be added if $t = 15° - n$.

Density of water at ordinary temperatures.—(*Rossetti*):

Temperature.	Densities.	Temperature.	Densities.
0°	·999871	15°	·99916
2°	·999928	20°	·998259
4°	1	25°	·99712
6°	·99997	30°	·995765
8°	·999886	100°	·95865
10°	·999747		

SPECIFIC GRAVITIES.

TABLE OF THE DEGREES OF BAUMÉ AND BECK
For Liquids Heavier than Water, with corresponding Specific Gravities.

Degrees Baumé or Beck.	Corresponding Specific Gravities.		Degrees Baumé or Beck.	Corresponding Specific Gravities.	
	Baumé.	Beck.		Baumé.	Beck.
0	1	1	37	1·3447	1·2782
1	1·0039	1·0059	38	1·3574	1·2879
2	1·014	1·0119	39	1·3703	1·2977
3	1·0212	1·018	40	1·3834	1·3077
4	1·0285	1·0241	41	1·3968	1·3178
5	1·0358	1·0303	42	1·4105	1·3281
6	1·0434	1·0366	43	1·4244	1·3386
7	1·0509	1·0429	44	1·4386	1·3492
8	1·0587	1·0494	45	1·4531	1·36
9	1·0665	1·0559	46	1·4678	1·371
10	1·0744	1·0625	47	1·4828	1·3821
11	1·0825	1·0692	48	1·4984	1·3934
12	1·0907	1·0759	49	1·5141	1·405
13	1·099	1·0828	50	1·5301	1·4167
14	1·1074	1·0897	51	1·5466	1·4286
15	1·116	1·0968	52	1·5633	1·4407
16	1·1247	1·1039	53	1·5804	1·453
17	1·1335	1·1111	54	1·5978	1·4655
18	1·1425	1·1184	55	1·6158	1·4783
19	1·1516	1·1258	56	1·6342	1·4912
20	1·1608	1·1333	57	1·6529	1·5044
21	1·1702	1·1409	58	1·672	1·5179
22	1·1798	1·1486	59	1·6916	1·5315
23	1·1896	1·1565	60	1·7116	1·5454
24	1.1994	1·1644	61	1·7322	1·5596
25	1·2095	1·1724	62	1·7532	1·5741
26	1·2198	1·1806	63	1·7748	1·5888
27	1·2301	1·1888	64	1·7969	1·6038
28	1·2407	1·1972	65	1·8195	1·619
29	1·2515	1·2057	66	1·8428	1·6346
30	1·2624	1·2143	67	1·839	1·6505
31	1·2736	1·2230	68	1·864	1·6667
32	1·2849	1·2319	69	1·885	1·6832
33	1·2965	1·2409	70	1·909	1·7
34	1·3082	1·25	71	1·935	
35	1·3202	1·2593	72	1·96	
36	1·3324	1·268			

SPECIFIC GRAVITIES

Of Solutions of Sulphuric Acid in Water at 15° C. (J. Kolb).

Degrees, Baumé.	Specific Gravities.	100 Parts by Weight Contain			Degrees, Baumé.	Specific Gravities.	100 Parts by Weight Contain		
		SO_3 p. 100.	H_2SO_4 p. 100.	Acid at 60° Baumé.			SO_3 p. 100.	H_2SO_4 p. 100.	Acid at 60° Baumé.
0	1	·7	·9	1·2	34	1·308	32·8	40·2	51·1
1	1·007	1·5	1·9	2·4	35	1·32	33·9	41·6	53·3
2	1·014	2·3	2·8	3·6	36	1·332	35·1	43	55·1
3	1·022	3·1	3·8	4·9	37	1·345	36·2	44·4	56·9
4	1·029	3·9	4·8	6·1	38	1·357	37·2	45·5	58·3
5	1·037	4·7	5·8	7·4	39	1·37	38·3	46·9	60
6	1·045	5·6	6·8	8·7	40	1·383	39·5	48·3	61·9
7	1·052	6·4	7·8	10					
8	1·06	7·2	8·8	11·3	41	1·397	40·7	49·8	63·8
9	1·067	8	9·8	12·6	42	1·41	41·8	51·2	65·6
10	1·075	8·8	10·8	13·8	43	1·424	42·9	52·8	67·4
					44	1·438	44·1	54	69·1
11	1·083	9·7	11·9	15·2	45	1·453	45·2	55·4	70·9
12	1·091	10·6	13	16·7	46	1·468	46·4	56·9	72·9
13	1·1	11·5	14·1	18·1	47	1·483	47·6	58·3	74·7
14	1·108	12·4	15·2	19·5	48	1·498	48·7	59·6	76·3
15	1·116	13·2	16·2	20·7	49	1·514	49·8	61	78·1
16	1·125	14·1	17·3	22·2	50	1·530	51	62·5	80
17	1·134	15·1	18·5	23·7					
18	1·142	16	19·6	25·1	51	1·540	52·2	64	82
19	1·152	17	20·8	26·6	52	1·563	53·5	65·5	83·9
20	1·162	18	22·2	28·4	53	1·58	54·9	67·	85·8
					54	1·597	56	68·6	87·8
21	1·171	19	23·3	29·8	55	1·615	57·1	70·	89·6
22	1·18	20	24·5	31·4	56	1·634	58·4	71·6	91·7
23	1·19	21·1	25·8	33	57	1·652	59·7	73·2	93·7
24	1·2	22·1	27·1	34·7	58	1·672	61	74·7	95·7
25	1·21	23·2	28·4	36·4	59	1·691	62·4	76·4	97·8
26	1·22	24·2	29·6	37·9	60	1·711	63·8	78·1	100
27	1·231	25·3	31	39·7					
28	1·241	26·3	32·2	41·2	61	1·732	65·2	79·9	102·3
29	1·252	27·3	33·4	42·8	62	1·753	66·7	81·7	104·6
30	1·263	28·3	34·7	44·4	63	1·774	68·7	84·1	107·7
					64	1·796	70·6	86·5	110·8
31	1·274	29·4	36	46·1	65	1·819	73·2	89·7	114·8
32	1·285	30·5	37·4	47·9	66	1·842	81·6	100	128
33	1·297	31·7	38·8	49·7					

DENSITIES OF SOLUTIONS OF NITRIC ACID AT 15° C.,
Giving the Percentage of Nitric Acid HNO_3 or Nitric Anhydride N_2O_5.

Densities.	Degrees Baumé.	Composition.	Water per Cent.	Nitric Acid HNO_3 per Cent.	Nitric Anhydride N_2O_5 per Cent.	Boiling Point.
1·522	49·3	HNO_3	0	100	85·8	86°
1·486	46·5	+ 1/2 H_2O	11·25	88·75	75·1	99
1·452	45	H_2O	22·22	77·78	66·7	115
1·42	42·6	3/2 H_2O	30	70	60·1	123
1·39	40·4	2 H_2O	36·36	63·64	54·5	119
1·361	35·2	5/2 H_2O	41·67	58·33	50·1	117
1·338	36·5	3 H_2O	46·16	53·84	46·2	117
1.315	34·5	7/2 H_2O	50	50	42·9	113
1·297	33·2	4 H_2O	53·33	46·67	40·1	113
1·277	31·4	9/2 H_2O	56·25	43·75	37·6	113
1·26	29·7	5 H_2O	58·82	41·18	35·4	113
1·245	28·4	11/2 H_2O	61·11	38·89	33·4	113
1·232	27·2	6 H_2O	63·16	36·84	31·6	113
1·219	25·8	13/2 H_2O	65	35	30·1	113
1·207	24·7	7 H_2O	66·67	33·33	28·6	108
1·197	23·8	15/2 H_2O	68·18	31·82	27·3	108
1·188	22·9	8 H_2O	69·56	30·44	26·1	108
1·18	22	17/2 H_2O	70·83	29·17	25	108
1·173	21	9 H_2O	72	28	24	108
1·166	20·4	19/2 H_2O	73·08	26·92	23·1	108
1·16	19·9	10 H_2O	74·07	25·93	22·2	108
1·155	19·3	21/2 H_2O	75	25	21·4	about 104°

DENSITIES OF SOLUTIONS OF ZINC SULPHATE AT 15° C. (*Gerlach*).

$ZnSO_4$ per 100 by Weight.	Densities.	$ZnSO_4 + 7H_2O$ per 100 by Weight.	Densities.
5	1·0288	35	1·231
10	1·0593	40	1·2709
15	1·0905	45	1·31
20	1·1236	50	1·3522
25	1·1574	55	1·3986
30	1·1933	60	1·4451

DENSITIES OF SOLUTIONS OF COMMON SALT AT +15° C.

(Gerlach).

NaCl PER 100 BY WEIGHT.	DENSITIES.	NaCl PER 100 BY WEIGHT.	DENSITIES.
2	1·01450	16	1·11938
4	1·029	18	1·13523
6	1·04366	20	1·15107
8	1·05851	22	1·16755
10	1·07335	24	1·18404
12	1·08859	26	1·20098
14	1·10384	26·39.*	1·20433

* Saturation.

SPECIFIC GRAVITY OF GASES AND VAPOURS

(Berthelot).

NAMES.	DENSITY, AIR BEING 1.	WEIGHT IN GRAMMES OF A CUBIC DECIMÈTRE AT 0° C. AND 760 MM. PRESSURE.
Air	1	1·2932
Oxygen	1·056	1·43
Hydrogen	·03926	·08958
Nitrogen	·9714	1·256
Chlorine	2·47	3·18
Bromine	5·54	7·16
Iodine	8·716	11·3
Mercury	6·976	8·96
Ammonia	·597	·761
Oxide of carbon	·968	1·254
Carbonic acid	1·529	1·9774
Vapour of water	·6235	·806
,, of absolute alcohol	1·613	2·095
,, of sulphuric ether	2·586	3·395
,, of essence of turpentine	5·013	6·512

DENSITIES OF SOLUTIONS OF COPPER SULPHATE AT 15° C. (*Gerlach*).

Percentage by Weight of $CuSO_4 + 5H_2O$.	Densities.	Percentage by Weight of $CuSO_4 + 5H_2O$.	Densities.
2	1·0126	14	1·0923
4	1·0254	16	1·1063
6	1·0384	18	1·1208
8	1·0516	20	1·1354
10	1·0649	22	1·1501
12	1·0785	24	1·1659

BAROMETER.

Correct formula for reduction to 0° of the height of the barometer.

$$h = H \frac{5550}{5550 + t} (1 + Kt).$$

h, reduced height; H, observed height; t, temperature at the time of observation in degrees C.; K, linear coefficient of expansion of the scale.

For brass . . . . $K = ·000016782$.
,, crystal . . . . . . $K = ·00007567$.

MEAN HEIGHT OF BAROMETER FOR DIFFERENT HEIGHTS ABOVE THE SEA LEVEL.

Height.	Height of Barometer.	Height.	Height of Barometer.
mètres.	millimètres.	mètres.	millimètres.
0	762	1147	660
21	760	1269	650
127	750	1393	640
234	740	1519	630
342	730	1647	620
453	720	1777	610
564	710	1909	600
678	700	2043	590
793	690	2180	580
909	680	2318	570
1027	670	2460	560

MEASUREMENT OF TEMPERATURE.
FAHRENHEIT AND CENTIGRADE THERMOMETER SCALES.

Fahren.	Centigrade.	Fahren.	Centigrade.	Fahren.	Centigrade.
− 4°	− 20°	33°	·56	70°	21·11
− 3	− 19·44	34	1·11	71	21·67
− 2	− 18·89	35	1·67	72	22·22
− 1	− 18·33	36	2·22	73	22·78
0	− 17·78	37	2·78	74	23·33
1	− 17·22	38	3·33	75	23·89
2	− 16·67	39	3·89	76	24·44
3	− 16·11	40	4·44	77	25
4	− 15·56	41	5	78	25·56
5	− 15	42	5·56	79	26·11
6	− 14·44	43	6·11	80	26·67
7	− 13·89	44	6·67	81	27·22
8	− 13·33	45	7·22	82	27·78
9	− 12·78	46	7·78	83	28·33
10	− 12·22	47	8·33	84	28·89
11	− 11·67	48	8·89	85	29·44
12	− 11·11	49	9·44	86	30
13	− 10·56	50	10	87	30·56
14	− 10	51	10·56	88	31·11
15	− 9·44	52	11·11	89	31·67
16	− 8·89	53	11·67	90	32·22
17	− 8·33	54	12·22	91	32·78
18	− 7·78	55	12·78	92	33·33
19	− 7·22	56	13·33	93	33·89
20	− 6·67	57	13·89	94	34·44
21	− 6·11	58	14·44	95	35
22	− 5·56	59	15	96	35·56
23	− 5	60	15·56	97	36·11
24	− 4·44	61	16·11	98	36·67
25	− 3·89	62	16·67	99	37·22
26	− 3·33	63	17·22	100	37·78
27	− 2·78	64	17·78	101	38·33
28	− 2·22	65	18·33	102	38·89
29	− 1·67	66	18·89	103	39·44
30	− 1·11	67	19·44	104	40
31	− 0·53	68	20	105	40·53
32	0	69	20·56	106	41·11

DETERMINATION OF HIGH TEMPERATURES

In Degrees C. by the Colour of Hot Platinum (*Pouillet*).

Colour of Platinum.	Corresponding Temperature.	Colour of Platinum.	Corresponding Temperature.
Beginning to be red	525°	Dark orange	1100°
Dark red	700	Bright orange	1200
Beginning to be cherry-red	800	White	1300
Cherry-red	900	Welding white	1400
Bright cherry-red	1000	Dazzling white	1500

CUBIC COEFFICIENTS OF EXPANSION OF SOME SOLIDS

For 1° between 0° and 100° C.

Solid.	Coeffic.	Solid.	Coeffic
	0·0000		0·0000
Steel	11500	Granite	08625
Tempered steel	12250	Gypsum	14010
Aluminium	22259	White marble	10720
Silver	19097	Black marble	04260
Pine wood	03520	Gold	15136
Bricks	05502	Platinum	08842
Bronze	18492	Lead	28484
Pine charcoal	10000	Fluor spar	20700
Brass	18782	Glass tube	08969
Copper	17182	,, rod	09220
Tin	21730	Plain glass	08613
Wrought iron	11821	Plate glass (St. Gobain)	08909
Iron wire	14401	Flint glass	08167
Cast iron	11100	Zinc	29680
Ice from − 27 to − 1	51813		

CUBIC COEFFICIENT OF EXPANSION OF MERCURY.

Absolute between 0° and 100° K = $\frac{1}{5550}$ = 0·00018018o.

Apparent in glass: $\frac{1}{6480}$ = 0·0001544.

MELTING AND BOILING POINTS OF COMMON SUBSTANCES.

The Boiling Points Determined at the Pressure of 760 mm. of Mercury.

SUBSTANCES.	MELTING.	BOILING.
Absolute alcohol	$< -90°$	78·3°
Alloys (See Fusible alloys)	—	—
Aluminium	600	—
Amber	288	—
Antimony	440	—
Arsenic	210	—
Benzine	7	80·8
Bismuth	265	—
Bromine	− 7·5	63
Bronze	900	—
Butter	30	—
Cadmium	320	860
Cane sugar	160	—
Carbon bisulphide	—	48
Carbonic acid	—	− 78
Cast iron	1050 to 1200	—
Caustic potash (saturated solution)	—	175
Chloride of sodium (saturated solution)	—	108
Copper	1050	—
Distilled water	0	100
Essence of turpentine	− 10	156·8
Fine gold	1250	—
Gold $\frac{900}{1000}$ths	1180	—
Iodine	107	176
Lead	335	—
Linseed oil	− 20	387·5
Mercury	− 39·5	350
Nitrate of silver	198	—
Olive oil	2·5	—
Palm oil	29	—
Paraffin	43·7	370
Petroleum	—	106
Phosphorus	44·2	290
Platinum	2000	—
Sea water	− 2·5	103·7
Selenium	217	665
Silver	1000	—
Spermaceti	49	—
Stearic acid	70	—
Stearine	61	—
Steel	1300 to 1400	—
Sulphur	114·5	400
Sulphuric æther	− 32	35·5
Sulphurous acid	− 79·2	− 10
Tallow	33	—
White wax	76·2	—
Wrought iron	1500 to 1600	—
Yellow wax	68·7	—
Zinc	412	1040

Boiling points of liquids at pressure 760 mm. of mercury (*Regnault*).

	Degrees.		Degrees.
Absolute alcohol (?)	− 90	Chloroform	60·16
Æther	34·97	Essence of turpentine	159·15
Alcohol	78·26	Mercury	357·25
Ammonia	− 38·5	Nitrogen monoxide	− 87·9
Benzine	80·36	Sulphuric acid	− 61·8
Carbon bisulphide	46·20	Sulphurous acid	− 10·08
Carbonic acid	− 78·2	Water	100
Chlorine	− 33·6		

HEAT DISENGAGED BY THE OXYDATION OF 1 GRAMME (*Everett*).
In Calories (Gramme Degree).

SUBSTANCES.	COMPOUND FORMED.	HEAT DISENGAGED.
Alcohol	—	6,900
Carbon	CO_2	8,000
Carbon monoxide	CO_2	2,420
Copper	CuO	602
Hydrogen	H_2O	34,000
Iron	Fe_3O_4	1,576
Marsh gas	$CO_2 + H_2O$	13,100
Olefiant gas	—	11,900
Phosphorus	P_2O_5	5,747
Sulphur	SO_2	2,300
Tin	SnO_2	1,233
Zinc	ZnO	1,301

HEAT DISENGAGED BY THE COMBINATION OF 1 GRAMME WITH CHLORINE (*Everett*).

SUBSTANCES.	COMPOUND FORMED.	HEAT DISENGAGED.
Copper	$CuCl_2$	961
Hydrogen	HCl	23,000
Iron	Fe_2Cl_6	1,745
Potassium	KCl	2,655
Tin	$SnCl_4$	1,079
Zinc	$ZnCl_2$	1,529

Heat disengaged (+) or absorbed (−) by chemical actions (*Favre and Silbermann*). (The figures relate to the equivalent in grammes in relation to that of hydrogen taken as unity, and not to unity of weight.) They are expressed in calories (g.-d.).

Oxydation of amalgamated zinc	+ 42,800
Combination of zinc oxide with sulphuric acid	+ 10,450
Decomposition of water	− 34,450
Decomposition of copper sulphate	− 29,600
Decomposition of nitric acid:	
a. Nitrogen dioxide and oxygen	− 6,880
b. Nitrous acid and oxygen	− 13,650

Heat disengaged by combustion of common gas.—The complete combustion of one cubic mètre of gas (= 35·317 cubic feet) produces 8000 calories (kg.-d.). In practice it is reckoned in town burners at 5000 calories per cubic mètre (35·317 cubic feet) on account of the incomplete combustion. In the case of burners which burn 100 litres (= 3·5 cubic feet) per carcel the heat developed is 500 calories (kg.-d.) per hour per carcel. Powerful Siemens burners (40 litres (= 1·4 cubic feet) per carcel) disengage 200 calories (kg.-d.) per hour per carcel.

Heat developed by the electric light in calories (kg.-d.) per hour per carcel, neglecting the combustion of the carbons.

Small incandescent lamps	10
Small arc lights (candles)	20
Powerful arc lights	5

As one cheval-heure (nearly one horse-power) corresponds to 270,000 kgm. or 637 calories (kg.-d.) the above figures assume that the light produced per cheval or per horse-power of electrical energy is:

Small incandescent lights	16 carcels.
Small arc lights	32 ,,
Powerful arc lights	128 ,,

which corresponds tolerably to the mean figures given by experiment.

Heat of evaporation.—In calories (g.-d.) per gramme at a pressure of 760 mm. of mercury:

Water	537	Acetic acid	102
Methyl alcohol	264	Sulphuric æther	1
Common alcohol	208	Essence of turpentine	69

RESISTANCES.

Heat of fusion.—In calories (g.-d.) per gramme:

Water . . 79·25	Tin . . 14·25	Phosphorus . 5·03
Zinc . . 28·13	Sulphur . 9·37	Mercury . . 2·82

Specific heat.—*Solids and liquids:*
In calories (g.-d.) per gramme.

Silver . . . . ·057	Platinum . . . ·324
Copper . . . . ·0952	Lead . . . . ·0314
Tin . . . . ·0362	Sulphur . . . ·1776
Iron . . . . ·1138	Zinc . . . . ·0956
Mercury . . . ·0319	Ice . . . . ·504
Nickel . . . . ·1092	Water . . . 1
Gold . . . . ·0324	Alcohol . . . ·5475

Specific heat of gases at constant pressure.

Air . . . . ·2374	Ammonia . . . ·5084
Oxygen . . . . ·2175	Carbonic acid . . ·2169
Hydrogen . . . 3·4090	Sulphurous acid . ·1544

RESISTANCES.
LIST OF COMMON BODIES
In Order of Decreasing Conductivity or Increasing Resistance (*Culley*).

Conductors.	Semi-conductors.	Insulators or Dielectrics.
Silver.	Wood - charcoal and coke.	Wool.
Copper.		Silk.
Gold.	Acids.	Glass.†
Zinc.	Saline solutions.	Sealing wax.
Platinum.	Sea water.	Sulphur.
Iron.	Rarefied air.*	Resin.
Tin.	Melting ice.	Guttapercha.
Lead.	Pure water.	Indiarubber.
Mercury.	Stone.	Gum lac.
	Ice.	Paraffin.
	Dry wood.	Ebonite.
	Porcelain.	Dry air.
	Dry paper.	

* The place in this list of dry air depends on the degree of rarefaction.
† Certain kinds of glass when quite dry insulate better than guttapercha.

L

Resistance of metals and alloys.— The table below enables the calculation to be made of the resistance at 0° C. of a conductor of given material when its length and its sectional area (and its diameter, if the conductor be circular) or its length and its weight be known. It is only necessary to remember that the resistance of a conductor is proportional to its length, inversely proportional to its section, and inversely proportional to its weight. One of the three coefficients a a' or a'' is used, taking care to express lengths, sectional areas, diameters, and weights in suitable units.

RESISTANCE OF COMMON METALS AND ALLOYS AT 0° C.
(Matthiessen).

METAL.	Resistance of a cube one centimètre in side between two opposite faces (specific resistance). (a)	Resistance of a wire one mètre long and one millimètre in diameter. (a')	Resistance of a wire one mètre long weighing 1 gramme. (a'')	Percentage increase of resistance per degree centigrade.
	Microhms.	Ohms.	Ohms.	Ohms.
Annealed silver	1·521	·01937	·1544	·377
Hard silver	1·652	·02103	·168	—
Annealed copper	1·616	·02057	·144	·388
Hard copper	1·652	·02104	·1469	—
Annealed gold	2·081	·02650	·408	·365
Hard gold	2·118	·02697	·415	—
Annealed aluminium	2·945	·03751	·0757	—
Compressed zinc	5·689	·07244	·4007	·335
Annealed platinum	9·158	·1166	1·96	—
,, iron	9·825	·1251	·7654	·63
,, nickel	12·60	·1604	1·071	—
Compressed tin	13·36	·1701	·9738	·365
,, lead	19·85	·2526	2·257	·387
,, antimony	35·90	·4571	2·411	·389
,, bismuth	132·7	1·689	13·03	·354
Liquid mercury	99·74	1·2247	13·06	·072
2 silver, 1 platinum	24·66	·314	2·959	·031
German silver	21·17	·2695	1·85	·044
2 gold, 1 silver	10·99	·1399	1·668	·065

RELATIVE CONDUCTIVITY OF COPPER

Alloyed with Foreign Substances.—Pure Copper = 100 (*Matthiessen*).

SUBSTANCES Forming the Alloy with Pure Copper.	Conductivity compared with Pure Copper.	Temperature in Degrees C.
		°
0·5 % of carbon	77·87	18·3
0·18 ,, of sulphur	92·03	19·4
0·13 ,, of phosphorus	70·34	20
0·95 ,, ,,	24·16	22·1
2·5 ,, ,,	7·52	17·5
Copper with traces of arsenic	60·08	19·7
2·8 % of arsenic	13·66	19·3
5·4 ,, ,,	6·42	16·8
Copper with traces of zinc	88·41	19·
1·6 % of zinc	79·37	16·8
3·2 ,, ,,	59·23	10·3
0·48 ,, of iron	35·92	11·2
1·66 ,, ,,	28·01	13·1
1·33 ,, of tin	50·44	16·8
2·52 ,, ,,	33·93	17·1
4·9 ,, ,,	20·24	14·4
1·22 ,, of silver	90·34	20·7
2·45 ,, ,,	82·52	19·7
3·5 ,, of gold	67·94	18·1
10·0 ,, of aluminium	12·68	14

CONDUCTIVITY OF DIFFERENT SAMPLES OF COPPER
(*Matthiessen*).

SUBSTANCES.	Conductivity.	Temperature.
		°
Pure copper	100	0
Spanish copper (Rio Tinto)	14·24	14·8
Russian copper	59·34	12·7
Australian copper	88·86	14·
American copper	92·57	15·
Polished copper wire	72·27	15·7
Hard copper wire	71·03	17·3

Influence of temperature on the resistance of metals

(*Matthiessen*).—The resistance of metals increases with the temperature according to the following empirical formula:

$$R = r(1 + at + bt^2).$$

R = resistance at temperature t.
r = resistance at 0° C.
t = temperature in degrees C.
a and b are numerical coefficients.

The following are some of the values of a and b:

	a	b
Very pure metals	+ ·003824	+ ·00000126
Mercury	·0007485	− ·000000398
German silver	·0004433	+ ·000000152
Platinum silver alloy	·00031	—
Gold and silver alloy	·0006999	− ·000000062

CONDUCTIVITY OF METALS.

Coefficients for temperature t in degrees C.

METALS.	COEFFICIENTS.
Silver	$c = 100 - ·38278\,t + ·0009848\,t^2$
Copper	$c = 100 - ·38701\,t + ·0009009\,t^2$
Gold	$c = 100 - ·36745\,t + ·0008443\,t^2$
Zinc	$c = 100 - ·37047\,t + ·0008274\,t^2$
Cadmium	$c = 100 - ·36871\,t + ·0007575\,t^2$
Tin	$c = 100 - ·36029\,t + ·0006136\,t^2$
Lead	$c = 100 - ·38756\,t + ·0009146\,t^2$
Arsenic	$c = 100 - ·38996\,t + ·0008879\,t^2$
Antimony	$c = 100 - ·39826\,t + ·0010364\,t^2$
Bismuth	$c = 100 - ·35216\,t + ·0005728\,t^2$
Mean	$c = 100 - ·37647\,t + ·0008340\,t^2$

Influence of temperature on the resistance and conductivity of pure copper.

—Increase of temperature diminishes the conductivity and increases the resistance of copper. Between 0° and 100° this variation is about $\frac{38}{10000}$ per degree centigrade.

The annexed table gives the coefficients by which the resistance of pure copper wire must be multiplied in order to obtain its resistance at t degrees between 0° and 30°, and the corresponding figures for the conductivity table.

RESISTANCE AND CONDUCTIVITY OF PURE COPPER AT DIFFERENT TEMPERATURES.

Centigrade Temperature.	Resistance.	Conductivity.	Centigrade Temperature.	Resistance.	Conductivity.
0°	1·	1·	16°	1·06168	·9419
1	1·00381	·99624	17	1·06563	·93841
2	1·00756	·9925	18	1·06959	·93494
3	1·01135	·98878	19	1·07356	·93148
4	1·01515	·98508	20	1·07742	·92814
5	1·01896	·98139	21	1·08164	·92452
6	1·0228	·97771	22	1·08553	·92121
7	1·02663	·97403	23	1·08954	·91782
8	1·03048	·97042	24	1·09356	·91445
9	1·03435	·96679	25	1·09763	·9111
10	1·03822	·96319	26	1·10161	·90776
11	1·04199	·9597	27	1·10567	·90443
12	1·04599	·95603	28	1·11972	·90113
13	1·0499	·95247	29	1·11382	·89784
14	1·05406	·94893	30	1·11782	·89457
15	1·05774	·94541			

Electric light carbons (*M. Joubert's* experiments).

Carré's Carbons.—Specific resistance: 3927 microhms at 20° C.

RESISTANCE OF CYLINDRICAL CARBONS PER MÈTRE.

Diameter in Millimètres.	Resistance in Ohms.	Diameter in Millimètres.	Resistance in Ohms.
1	50	8	·781
2	12·5	10	·5
3	5·55	12	·348
4	3·125	15	·222
5	2·	18	·154
6	1·39	20	·125

The resistance diminishes as the temperature increases. Between 0° and 100° C. the coefficient of reduction is $\frac{1}{1912}$ per degree centigrade.

Retort carbon.—Specific resistance 66750 microhms, about.

Graphite.—Very variable, between 2400 and 42000 microhms.

Gaudin's carbon (Mignon and Rouart).—Specific resistance 8513 microhms. Between 0° and 100° C., the resistance diminishes $\frac{1}{24}$th.

Coating of carbons with metal under ordinary conditions reduces their resistance to $\frac{1}{3}$ of its original value.

Metalloids.—*Crystallised selenium.* Specific resistance at 100° C., 60,000 ohms.

Red phosphorus.—132 ohms at 20°.

Tellurium.—·213 ohms at 20°.

Specific resistance of some liquids at 14° and 24° in ohms per cubic centimètre (*Blavier*).

	14°	24°
Solution of copper sulphate (8 per cent.)	45·7	37·1
,, ,, (28 per cent.)	24·7	18·8
Saturated solution of zinc sulphate.	21·5	17·8
Solution of sulphuric acid (density = 1·1)	·88	·73
,, ,, (density = 1·7)	4·67	3·07
Nitric acid (density = 1·36)	1·45	1·22
Distilled water (*Pouillet*), unknown temperature		932
Distilled water with $\frac{1}{20000}$ of sulphuric acid		1550

CONDUCTIVITY OF SOLUTIONS.

Pure Copper = 100,000,000 (*Matthiessen.*)

SOLUTIONS.	Centigrade Temperature	Conductivity.
1. Copper sulphate concentrated	9°	5·42
,, ,, with an equal volume of water	,,	3·47
,, ,, with 3 volumes of water	,,	2·08
2. Common salt concentrated	13°	31·52
,, ,, with an equal volume of water	,,	23·08
,, ,, with 2 volumes of water	,,	17·48
,, ,, with 3 volumes of water	,,	13·58
3. Zinc sulphate concentrated	14°	5·77
,, ,, with 1 volume of water	,,	7·1
,, ,, with 3 volumes of water	,,	5·43

SPECIFIC RESISTANCE OF SOLUTIONS OF SULPHURIC ACID.
(Matthiessen).

Specific Gravity.	Percentage by weight of Sulphuric Acid.	Centigrade Temperature.	Resistance.
1·003	·5	16·1	16·01
1·018	2·2	15·2	5·47
1·053	7·9	13·7	1·884
1·080	12	12·8	1·368
1·147	20·8	13·6	·96
1·190	26·4	13	·871
1·215	29·6	12·3	·83
1·225	30·9	13·6	·862
1·252	34·3	13·5	·874
1·277	37·3	—	·900
1·348	45·4	17·9	·973
1·393	50·5	14·5	1·086
1·493	60·6	13·8	1·549
1·638	73·7	14·3	2·786
1·726	81·2	16·3	4·337
1·827	92·7	14·3	5·32

SPECIFIC RESISTANCE OF NITRIC ACID (D = 1·36).
(Temperature in Degrees C.)

2°	1·94	8°	1·65	16°	1·39	24°	1·22
4°	1·83	12°	1·5	20°	1·3	28°	1·18

SPECIFIC RESISTANCE OF SOLUTIONS OF COPPER SULPHATE AT 10° C. *(Ewing and MacGregor).*

Density.	Specific Resistance.	Density.	Specific Resistance.
1·0167	164·4	1·1386	35
1·0216	134·8	1·1432	34·1
1·0318	98·7	1·1679	31·7
1·0622	59	1·1823	30·6
1·0858	47·3	1·2051 (saturated)	29·3
1·1174	38·1		

SPECIFIC RESISTANCE OF SOLUTIONS OF ZINC SULPHATE AT 10° C. (*Ewing and MacGregor*).

Density.	Specific Resistance.	Density.	Specific Resistance.
1·014	182·9	1·2709	28·5
1·0187	140·5	1·2891	28·3 (minimum)
1·0278	111·1	1·2895	28·5
1·054	63·8	1·2987	28·7
1·076	50·8	1·3288	29·2
1·1019	42·1	1·353	31·
1·1582	33·7	1·4053	32·1
1·1845	32·1	1·4174	33·4
1·2186	30·3	1·422 (saturated)	33·7
1·2562	29·2		

The above table shows that the most saturated solution is not always the best conductor; the same phenomenon is observed with a solution of common salt.

Mixtures of zinc sulphate and copper sulphate.

—The resistance of this mixture is always less than the mean of the resistances of the two solutions, often it is less than the resistance of the least resisting solution (*Ewing* and *MacGregor*).

APPROXIMATE SPECIFIC RESISTANCES OF WATER AND ICE
(*Ayrton and Perry*).

Centigrade Temperature.	Specific Resistance in Megohms.	Centigrade Temperature.	Specific Resistance in Megohms.
− 12·4	2240	− ·2	284
− 6·2	1023	+ ·75	118·8
− 5·02	948·6	+ 2·2	24·8
− 3·5	642·8	+ 4	9·1
− 3·0	569·3	+ 7·75	·54
− 2·46	484·4	+ 11·02	·34
− 1·5	387·6		

Specific resistance of glass (*G. Foussereau*, 1882).
Common glass, soda and lime, $d = 2539$.

Temperature.	Specific resistance in millions of megohms.
$+ 61 \cdot 2°$	$\cdot 705$
$+ 20°$	91
$- 17°$	7970

Hard Bohemian glass, $d = 2 \cdot 431$. It has 10 to 15 times less resistance than common glass at the same temperature.

Flint glass, $d = 2 \cdot 933$. It is from 1000 to 1500 times better insulator than common glass at the same temperature. Its conductivity only begins to be manifested above $40°$ C.

At $46°$ specific resistance . . . = 6182 million megohms.
„ $105°$ „ „ . . . = 11·6 „ „

Approximate specific resistance of insulators after several minutes of electrification (*Ayrton and Perry*).

	Specific resistance in megohms.	Centigrade Temperature.
Mica	84×10^6	$20°$
Guttapercha	450×10^6	$24°$
Gum lac	$9,000 \times 10^6$	$28°$
Hooper's compound	$15,000 \times 10^6$	$24°$
Ebonite	$28,000 \times 10^6$	$46°$
Paraffin	$34,000 \times 10^6$	$46°$
Glass	Not accurately measured, but greater than the preceding.	
Air	Practically infinite.	

Specific resistance of guttapercha.—Varies from 1 to 20, according to the quality and degree of purification. It depends in one and the same sample on the temperature, the time of electrification, and the external pressure.

For certain samples at $24°$ C., the specific resistance $= 25 \times 10^{12}$ ohms.
For the best quality at $24°$ C., its specific resistance $= 500 \times 10^{12}$ ohms.

Generally manufacturers are required to supply at the specific resistance of 200×10^{12} ohms. This figure is often very much exceeded by good manufacture.

Influence of temperature (*Clark and Bright's* empirical formula):

$$R = R_0 a^t.$$

R specific resistance at temperature t.
R_0 specific resistance at $0°$ C.
a coefficient of which the mean value is ·8944.

Influence of length of time of electrification.—The specific resistance increases with the duration of the current which traverses the guttapercha. The variation is less the higher the temperature. The following table shows the influence of these two factors.

Minutes of Electrification	Resistance at 0°.	Resistance at 24°.	Minutes of Electrification	Resistance at 0°.	Resistance at 24°.
1	100·	5·51	20	230·8	7·33
2	127·9	6·	30	250·6	7·44
5	163·1	6·66	60	290·4	7·6
10	190·9	6·94	90	318·3	7·66

In cable testing, *one minute* is generally taken as the time of electrification.

Influence of pressure.—Pressure increases the insulation according to the following empirical formula:

$$R_p = R\,(1 + \cdot 00327\,p).$$

R_p specific resistance at pressure p.
R specific resistance at atmospheric pressure.
p pressure in kilogrammes per square centimètre, or in atmospheres.

At a depth of 4000 mètres, the specific resistance is more than doubled. The coefficient varies with the quality, and increases with the length of time the guttapercha has been in the water.

Specific resistance of indiarubber.

— Vulcanised rubber, or Hooper's compound, has been used sometimes for cables. Its resistance is greater than that of guttapercha, varies less with temperature; but indiarubber attacks the copper, which ought to be tinned. The insulator must also be applied, not in a pasty state like guttapercha, but in strips melted together, which complicates the process of manufacture.

Specific resistance at 0° . . . . = $32{,}000 \times 10^{12}$ ohms.
,,   ,,   24° . . . . = $7{,}500 \times 10^{12}$,,

CONDUCTORS.

Nature of conductors.—*Copper* is the metal most employed in applications of electricity. Coils of electrical instruments of all kinds and of all power are almost exclusively composed of it. Also the cores of subterranean and submarine cables. Electric light leads and branch leads in houses, etc., etc.

Iron is generally used for overhead telegraph lines. *Steel, phosphor-bronze,* and *silicium-bronze* for overhead telephone lines. Besides these generally used metals and alloys, the following metals are sometimes used because of their special qualities satisfying some particular want.

Silver.—For delicate instruments of high conductivity.

German silver.—For resistances, on account of its being little affected by temperature, and its high specific resistance.

Platinum silver alloy.—For the same reason.

Platinum.—Because it does not oxydise, its great resistance, and its high melting point.

Mercury.—For standards of resistance, because of its homogeneity after purification; it is most used, however, for making connections by means of mercury cups.

Aluminium.—For certain movable coils, has the advantage of being the best conductor of all metals for equal weight per unit of length.

Bare conductors.—Bare copper wires are from four hundredths of a millimètre to five millimètres in diameter; for larger sections, twisted cord of smaller wire is preferred. Sometimes copper is also used in the form of ribands, or prisms, of which the section is the segment of a circle.

Covered wire.—Bare wires are but little used in instruments. They are generally covered with an insulator. This insulator is generally a layer of cotton or of silk, single or double, soaked after the coils are wound, in paraffin, cement, or a special insulating varnish. (*See* Fifth Part.) In machines, wires are covered with bitumen or gum lac. When used as transmitting conductors in telegraphy, telephony, light, transmission of energy, etc., they are covered during the manufacture with more or less complicated insulators, and are then called *cables*.

The cables are *sheathed* or not, according as they are covered or not with a *sheathing* of iron or steel wire intended to protect them and give them the necessary strength to resist the traction efforts to which they are submitted, particularly in submarine cables.

DIAMETER OF CONDUCTING WIRES.

Gauges.—The wires employed in electricity are often known by the number of some particular gauge. In France up to the present time the decimal gauge has been used for wires of large diameter, and the *jauge carcasse* for fine wires, and the Limoge gauge for iron wire.

In England the Birmingham wire gauge is most commonly represented by the letters B.W.G.

In America there is the American gauge; in India, the Indian gauge, etc.

The numbering of wires ought completely to disappear and be replaced by a more rational and accurate system, the diameter expressed in millimètres or 100ths of a millimètre. Numbering only creates confusion and embarrassment, for all authors are not agreed on the diameter of a given number. This confusion is very great in England, where there are no less than 14 different Birmingham gauges. We will give here for reference the most probable diameter of the wires of the *jauge carcasse* and the Birmingham gauge in millimètres.

To avoid confusion, when we write the number of a wire we will always follow this number by the real diameter expressed in millimètres. A new gauge has been introduced in England, but it is not found to answer, and is never used in specifications.

BIRMINGHAM GAUGE.—B. W. G. (*Hotzapffel*).

B.W.G. Number.	Diameter in Millimètres.	B.W.G. Number.	Diameter in Millimètres.	B.W.G. Number.	Diameter in Millimètres.
0000	11·531	11	3·048	25	·508
000	10·795	12	2·769	26	·457
00	9·652	13	2·413	27	·406
0	8·636	14	2·108	28	·356
1	7·62	15	1·829	29	·330
2	7·213	16	1·651	30	·305
3	6·579	17	1·473	31	·254
4	6·045	18	1·245	32	·229
5	5·588	19	1·067	33	·203
6	5·154	20	·889	34	·178
7	4·572	21	·813	35	·127
8	4·191	22	·711	36	·102
9	3·759	23	·635		
10	3·404	24	·559		

JAUGE CARCASSE.
Approximate Diameters in Hundredths of a Millimètre.

Number.	Diameter.	Number.	Diameter.	Number.	Diameter.
P	50	24	29	38	11
12	47	26	26	40	10
14	44	8	22	42	9
16	0	30	20	44	8
18	37	32	17	46	7
20	34	34	14	48	6
22	32	36	12	50	5

Copper.—Specific gravity, 8·878.

Breaking strain, 26·7 kg. per square millimètre.

Weight per kilomètre, 6·973 d^2 kilogrammes (d diameter in millimètres).

Diameter of a wire weighing n kilog. per kilomètre $= \sqrt{n} \times \cdot 1434$.

The resistance per kilomètre of pure copper wire $= \frac{21 \cdot 84}{d^2}$.

A pure copper wire weighing one kilogramme per kilomètre has a resistance of
$$143 \cdot 85 \text{ ohms at } 0° \text{ C.}; \quad 152 \cdot 5 \text{ at } 15 \cdot 5° \text{ C.}$$

Pure copper wire weighing n grammes l mètres long has a resistance of

$$\frac{\cdot 144 \times l^2}{n} \text{ ohms at } 0° \text{ C.}$$

$$\frac{\cdot 1525 \times l^2}{n} \text{ ohms at } 15 \cdot 5° \text{ C.}$$

The resistance increases ·388 per cent. per degree centigrade, or ·215 per cent. per degree Fahrenheit.

A pure copper wire weighing one kilogramme per nautical mile or knot (1852 mètres) at 24° C. has a resistance of 540·8 ohms.

A pure copper wire weighing one kilogramme per nautical mile at 24° C. has a resistance of 291·54 ohms per kilomètre.

Resistance of a commercial copper wire of conductivity c, that of pure copper being taken as equal to 1, is given by the formula,

$$R_c = R_p \times \frac{1}{c}.$$

R_c being the resistance of the commercial wire, and R_p that of a pure copper wire of the same dimensions at 0° C.

RESISTANCE OF WIRES OF PURE ANNEALED COPPER AT 0° C.
(Density = 8·9.)

Number in the Decimal Gauge.	Diameter in Millimètres.	Weight per Mètre in Grammes.	Length in Mètres per Kilogramme (Bare Wire).	Resistance of Wire of Pure Annealed Copper at 0° C.		
				Ohms per Kilomètre.	Mètres per Ohm.	Ohms per Kilogramme.
—	5	175	5·7	·8	1230·5	·00456
20	4·4	135·28	7·4	1·03	944·38	·00784
19	3·9	106·35	9·5	1·35	722	·0128
18	3·4	80·8	12·5	1·60	563·92	·0222
17	3	62·93	16	2·3	439·07	·0365
16	2·7	51	19·8	2·8	355·65	·0557
15	2·4	40·23	25	3·6	281	·088
14	2·2	33·82	29	4·2	236·08	·123
13	2	27·95	36	5·1	195·15	·185
12	1·8	22·7	44	6·3	158·08	·278
11	1·6	17·89	56	8	124·9	·448
10	1·5	15·75	63	9·1	109·75	·574
9	1·4	13·7	73	10·5	95·651	·763
8	1·3	11·84	85	12	82·42	1·03
7	1·2	10·06	100	14	70·247	1·42
6	1·1	8·47	119	17	59·024	2·02
5	1	6·99	144	20	48·782	2·95
4	·9	5·66	173	25	39·515	4·19
3	·8	4·47	225	32	31·225	7·21
2	·7	2·83	294	42	23·9	12·3
1	·6	2·52	400	57	17·56	22·78
P	·5	1·74	576	81	12·305	46·81
—	·4	1·175	902	122·4	8·173	110·41
—	·34	·808	1251	177·9	5·622	222·55
—	·3	·7181	1607	228·5	4·377	367·2
—	·24	·4026	2508	357	2·801	895·36
—	·2	·2797	3614	514	1·945	1,857·6
—	·16	·179	5590	803·1	1·245	4,489
—	·12	·1007	9929	1428	·7	14,179
—	·1	·0699	14369	2056	·486	29,549
—	·08	·0447	24570	3213	·311	78,943
—	·06	·0252	39824	5713	·173	227,515
—	·04	·0112	88878	12848	·078	1,142,405

Copper wire one millimètre in diameter.

Section	·7854 sq. mm.
Weight of one mètre	6·99 grammes.
Resistance of annealed copper wire:	
At 0° (pure)	·02057 ohm.
Hard drawn, at 0° (pure)	·02104 ,,
Annealed at 15° (pure)	·021767 ,,
Hard drawn at 15° (pure)	·022263 ,,
Annealed at 15° (conductivity = ·9)	·024185 ,,
Hard drawn at 15° (conductivity = ·9)	·024707 ,,
Number of mètres per kilogramme	144 mètres.
Resistance per kilogramme of pure copper wire at 0°	2·95 ohms.
Weight per kilomètre	6·973 kilogrammes.

Approximate weight of the silk covering of wires per kilogramme (*Culley*).

Diameter of bare wire.	Number of mètres per kilogramme.	Weight of silk in grammes.
1·6	57	34
1	140	51
·66	328	68
·35	1140	102
·22	3000	136
·18	4500	187
·13	8800	

Iron.—Specific gravity, 7·79.

Breaking strain, 40 kilogrammes per square millimètre (annealed).

Breaking strain, 60 kilogrammes per square millimètre (not annealed).

Mean weight of wire 4 millimètres diameter, 100 kilog. per kilomètre.

Electric resistance at 0°: 9 ohms per kilomètre.

Wire of diameter d, $R = \frac{144}{d^2}$ ohms per kilomètre.

Increase of resistance with temperature, 0·0063 per degree centigrade.

Resistance of steel wires, 1·28 that of galvanised iron.

Galvanised wire.—(*Specification of the French telegraphs*). Charcoal iron annealed wire, 5 millimètres in diameter, ought to carry a weight of 650 kilogs.; 4 millimètres, 440; 3-millimètre wire, 350. Permanent elongation under this strain ought not to exceed 6 per cent. of the length.

The wire must be rolled on a cylinder and submitted to a tension of 500 kilogs. for 5 millimètres; for the 4-millimètre wire, 200; for the 3-millimètre wire, 200; without breaking or exceeding the limit of elongation. It must be bent at right angles first in one direction and then in the other, three times for the wire of 5 millimètres, and four times for the wire of 4 millimètres, and five times for the 3-millimètre wire.

To test the galvanisation, the wire must bear, without the iron becoming exposed even partially, four successive immersions of one minute each in a solution of sulphate of copper in five times its weight of water.

The wire must bear being rolled round a cylinder one centimètre in diameter, without the layer of zinc cracking or scaling off. The wire must not be spotted with rust.

The tests are made on five coils out of 100, and the whole delivery is rejected if one-tenth of the wire does not satisfy this test.

The 5-millimètre wire is delivered in pieces of 25 kilogs. (or 160 mètres), the 4-millimètre wire in pieces of 25 kilogs. (250 mètres); 3-millimètre in pieces of 15 kilogs. (270 mètres).

The weight of zinc in well galvanised iron is 170 gr. per square mètre, or 2 kilogs. per mètre of 4-millimètre wire.

Phosphor bronze (*Lazare & Weiller*).—An alloy of copper and tin, in which phosphorus only plays a transitory part during the manufacture. Used for telegraphic and telephonic lines. Wire is used from ·8 to 1·1 millimètre in diameter, of which the weight per kilomètre varies from 4·5 to 7 kilogs. The spans may be from 400 to 500 mètres.

Silicium bronze (*Lazare & Weiller*).—Analogous to phosphor bronze, silicon replacing the phosphorus, it is a better conductor and more tenacious.

According to the purpose for which it is intended, silicium bronze can be prepared either of high conductivity or of great mechanical strength. Two examples may be given.

Telegraph wire.—2 mm. in diameter, breaking strain of 45 kg. per square mm., and conductivity 95 per cent. Weight, 28 kg. per kilomètre, and 5·43 ohms per kilomètre resistance at 0° C.

Telephone wire.—1·1 mm. in diameter, breaking strain 75 kg. per square mm., and conductivity 34 per cent. One kilomètre weighs 8·5 kg. and has a resistance of 57 ohms at 0° C. (*H. Vivarez*).

PROPERTIES OF PHOSPHOR AND SILICIUM BRONZE.

PROPERTIES.	BRONZE.	
	PHOSPHOR.	SILICIUM.
Breaking strain in kilogrammes per square millimètre	90	70
Breaking strain of a wire one millimètre in diameter	70·2	55·6
Resistance in ohms per kilomètre (wire 1 mm. in diameter)	60	34
Conductivity (pure copper = 100)	30	61
Weight of one kilomètre in kilogrammes (wire 1 mm. in diameter)	6·63	6·63

The conductivity of silicium bronze may be increased by diminishing its breaking strain. Thus, a wire, the breaking strain of which is 53 kg per square mm., has a conductivity of 88 per cent.

Conductors for commercial purposes.—They ought to be of as high conductivity as possible, and to have the greatest facility for cooling. The best insulator would be air, then vulcanised rubber, which is fairly diathermanous, and may be raised to 100° without accident. Guttapercha is inferior to rubber, and gets soft when it is hot, and allows the conductors to come into contact. The lead-covered cables remain dry, and are less exposed to accidents, but they do not coil easily; the insulating layers ought to be as thin as possible in order to facilitate coiling. A fault in the insulation of wire under water would produce flickering of the light, because of the disengagement of gas due to the decomposition of the water. It is as well to test the conductivity of conductors and their insulation daily.

Conductors for electric lighting and transmission of energy.—The *India Rubber and Gutta Percha Telegraph Works Company* manufacture a series of special conductors insulated with indiarubber. The following are a few of their patterns:

M

Number of Wires.	Diameter of each Wire in Millimètres.	Resistance per Kilomètre in Ohms at 15·5° C.	Number of Wires.	Diameter of each Wire in Millimètres.	Resistance per Kilomètre in Ohms at 15·5° C.
7	·9	3·93	12	1·6	·72
7	1·2	2·07	14	1·6	·6
7	1·6	1·21	19	1·6	·43
7	1·8	·95	19	1·8	·35

These conductors are manufactured with three different classes of insulation, each one being defined by the insulation in megohms per 1,000 yards (910 mètres) at temperature 15° C.

NAME OF INSULATION.	Without Lead.	Leaded.
	megohms.	megohms.
Class A.—Light insulation for conductors exposed to the air . . .	·5	3
Class B.—Medium insulation for cables run in dry places . . .	10	15
Class C.—High insulation for damp places, tunnels, etc. . . .	100	150

Mechanical tests for insulators (*R. Sabine*).—A conductor covered with its insulation ought to bear several successive bendings without cracking. The elasticity ought to be that of a spring or a cane, not that of varnish or of paste. A length of 2 or 3 feet is detached by splitting the insulation lengthways with a penknife, and removing the conductor. The insulator supported at one end ought to bear a certain definite weight at the other. Indiarubber breaks with a weight equal to 300 times its own weight. Guttapercha carries much more. The insulator detached from the conductor and placed on an anvil ought not to break when it is struck with a hammer. Good indiarubber recovers its shape immediately; guttapercha takes a longer time. When old both split and crumble under the hammer. Insulators which crack off or split under a blow ought to be rejected, as an accident might produce the same injurious effects, and spoil the insulation of the conductor.

Specific inductive capacities (*Fleeming Jenkin*):

Air	1	Paraffin	1·98
Resin	1·77	Pure indiarubber	2·8
Pitch	1·8	Hooper's compound	3·1
Yellow wax	1·86	W. Smith's guttapercha	3·4
Glass	1·9	Guttapercha	4·2
Sulphur	1·93	Mica	5
Gum lac	1·95		

Capacity of condensers of ordinary forms in electrostatic units.

In these formulæ r represents the radius, k the dielectric capacity of the insulator, and C the capacity of the condenser.

SPHERE: $C = r$.

TWO CONCENTRIC SPHERES OF RADII r and r': $C = k \dfrac{rr'}{r' - r}$.

CYLINDER of length l: $C \dfrac{l}{2 \log_e \frac{l}{r}}$.

($\log_e$ = Naperian log).

TWO CONCENTRIC CYLINDERS of length l: r' radius of external cylinder; r radius of internal cylinder.

$$C = k \dfrac{l}{2 \log_e \frac{r'}{r}}.$$

CIRCULAR DISC of radius r and negligable thickness:

$$C = \dfrac{2r}{\pi}.$$

TWO PARALLEL CIRCULAR DISCS of radius r and area S, thickness of dielectric b:

$$C = k \dfrac{r^2}{4b} = k \dfrac{S}{4\pi b}.$$

(This last formula is applicable to two parallel discs of any shape when their dimensions are large as compared with their distance from each other b.)

MAGNETS.

The **power of magnets** is estimated by the weight they will carry. The most powerful magnet known is the laminated magnet belonging to M. Jamin, which weighs 50 kilogrammes and carries 500; another of the same form, weighing 6 kilogrammes, carries 80, or 13·5 times its own weight. Some small magnets carry as much as 25 times their own weight. A horse-shoe magnet carries three or four times the weight which a straight bar of the same mass can carry.

Bernouilli's rule.—If w be the weight of a magnet, p the weight which it can carry, then

$$p = a \sqrt[3]{w}.$$

a is a constant depending on the quality of the steel and the method of magnetisation. For the best qualities of *Wetteren* of Haarlem a varies between 19·5 and 23. The steels of *Allevard* have also a very high constant.

Mr. Le Neve Foster has found that all steels of very high constant contain tungsten, and that the constant increases with the quantity of tungsten present.

Coefficient of induced magnetism.—The magnetic moment of a long bar placed in a uniform magnetic field is, *for a weak field*, proportional to its length l, its sectional area s, the intensity of the field H, and a numerical constant k, which is called the coefficient of induced magnetism. k is negative for diamagnetic bodies. The following are the mean values given by Barlow and Plücker for some substances.

Magnetic Substances.	k	Diamagnetic Substances.	k
Soft iron (wrought)	32·8	Water	$-10·65 \times 10^{-}$
Cast iron	23	Sulphuric acid (d. = 1·839)	$-6·8 \times 10^{-6}$
Soft steel	21·6	Mercury	$-33·5 \times 10^{-6}$
Hard steel	17·4	Phosphorus	$-18·3 \times 10^{-6}$
Nickel	15·3	Bismuth	$-25·0 \times 10^{-6}$
Cobalt	32·8		

These coefficients are only approximate, and change for every substance with the absolute value of H. For example, according to *Weber*, in a very weak magnetic field the value of k is five times greater for nickel than for iron.

Supersaturation of magnets.—A magnetised bar often takes up a magnetic supersaturation, which it gradually loses at a progressively decreasing rate until its magnetisation is reduced to its permanent value. In all experiments depending on the constancy of magnets they ought to be magnetised at least six months beforehand.

Professor Hughes finds that by smartly hammering the magnetised bar it can be brought to its lowest limit of magnetisation. He suggests the use of this method whenever a trustworthy permanent magnet of small power is required; and, also, as a test of the durability of permanent magnets for compasses and other purposes, those that fall below a certain intensity being rejected, and those that are retained being re-magnetised.

Influence of temperature on magnetism.—Magnetisation diminishes with increase of temperature, and conversely. The diminution becomes permanent if the temperature rises too high. A magnet heated to cherry-red loses its magnetism, and at this temperature soft iron ceases to be attracted by a magnet. For nickel the temperature at which this effect takes place, called *the magnetic limit*, is about 350°.

Temper. Compressed steel.—The coercitive force of steel is greater the more it is tempered. *M. Clemandot* (1882) gives coercitive force to untempered steel by compressing it in a hydraulic press. This *mechanical temper* allows of the steel being worked by the file, chisel, or in the lathe, either before or after magnetisation, which is a valuable property in the construction of a great number of instruments.

Experimental determination of the moment of inertia of a magnetised bar (*Gauss*).—The bar or needle is made to oscillate: let t be the time of one oscillation; two equal weights q are suspended at equal distances from the centre; let the first distance be a_1, the second distance a_2; let t_1 and t_2 be the respective times of one oscillation, then

$$I = \frac{t^2}{t_1^2 - t_2^2} \times 2qt^2.$$

To bring an oscillating magnet to rest.—By means of a small magnet so held as to repel the nearest end of the oscillating magnet. At the moment when the oscillating magnet is passing zero as it swings towards the observer the small magnet is sharply brought near it. The oscillating magnet stops, and at the moment when it begins to swing in the opposite direction the magnet in the hand is withdrawn. A magnetised steel lever or turnscrew does very well.

METHODS OF MAGNETISATION.

Single touch.—Stroke the bar to be magnetised from end to end with one pole of a natural or artificial magnet, repeating the strokes in the same direction. The pole at the end of the bar at which the strokes begin is of the same name as the pole used to stroke it with.

Separate touch.—Two magnetised bars of the same name are placed at the middle of the bar to be magnetised with poles of contrary name downwards, these bars should be inclined at about 30°; they are separated, stroking the magnet in opposite directions, replaced in the middle and again separated, and so on. This method is quicker than the preceding.

Double touch.—Two magnets fixed in an inclined position in a wooden frame are passed from end to end of the bar to be magnetised, always in the same direction. A more regular magnetisation is thus obtained. The effects produced by the two last methods are better and quicker if the bar to be magnetised is laid on two magnets with their opposite poles facing each other, these poles to be of the same name as the pole of the magnetising magnet on the same side (*Coulomb*).

These methods are now replaced by others depending on the action of currents, the simplest being to rub the bar to be magnetised over the poles of a powerful electro-magnet.

Elias' method.—A coil through which a current is passing is passed backwards and forwards over the bar to be magnetised. For a horse-shoe magnet two coils are used, which are moved together up and down the two branches. With strong currents a bar may be magnetised to saturation by one pass. The rapidity of the passes has no influence on the magnetisation; their number has greater influence as the current is weaker.

To magnetise a needle.—Place the needle on a horse-shoe magnet, putting the end which is to be a north pole on the south pole of the magnet, and *vice versa*; rub the needle against the magnet two or three times in the direction of its length.

Armatures of magnetised bars.—To preserve the magnetism of the bars they should be provided at their ends with soft iron armatures, or they may be placed with their opposite poles in contact. The opposite poles tend to preserve their reciprocal magnetism, poles of the same name to destroy it. The armatures ought always to be *slid* off, and not *pulled* off. The earth's magnetism tends to preserve the magnetism of a freely suspended needle.

TERRESTRIAL MAGNETISM.

Elements of the earth's magnetism, *Jan. 1st*, 1879.

Declination at Paris (west)	16° 56
Mean annual diminution	0° 9'
Mean inclination at Paris	65° 32' 6''
Horizontal component of the earth's magnetism at Paris (*in dynes*)	·19324
Total force (*in dynes*)	·46485

Elements of the earth's magnetism at the observatory at the Père Saint-Maur, *Jan. 30th*, 1883 (*Mascart*).

Inclination	65° 17'
Declination (west)	16° 33'
Horizontal component (*in dynes*)	·1932
Total intensity (*in dynes*)	·46485

ELECTRO-MAGNETS.

Laws of electro-magnets.—When a magnet is far from the point of *magnetic saturation*, the following laws are applicable. The magnetic strength is proportional to the strength of the magnetising current and the number of turns of wire in the coil. It is independent of the size and nature of the conductor and the diameter of the coil.

When account is taken of magnetic saturation, *Müller* gives the following rule. The magnetic strength m of an electro-magnet is proportional to the arc of which the tangent is the strength C of the magnetising current.

$$m = A \tan^{-1} C.$$

Maximum attraction (*Joule*).—The maximum attraction is 200 pounds per square inch, or 14,515 grammes per square centimètre, which corresponds to an intensity of magnetisation $\gamma = 1500$:

$$f = 2\pi S \gamma^2.$$

f being the maximum attraction, and S the area of the attracting surface.

The formula shows that the attraction of an electro-magnet depends only on the diameter of its core, the length is only of use in separating the poles, and thus preventing them from interfering with each other. Joule has constructed small electro-magnets which carry as much as 3,500 times their own weight.

Action of a bar of iron in a solenoid.—A bar of iron introduced into a solenoid makes the internal magnetic field about 33 times more intense by concentrating the lines of force near the poles, and allows the current to produce powerful effects in a limited space. Looked at in this way the action is analogous to that of a lens which concentrates a ray of light on to the point where the maximum luminous action is required (*Fleeming Jenkin*).

Formulæ for electro-magnets before magnetic saturation.—For relatively feeble currents and soft iron cores of which the length is much greater than the diameter, such as the electro-magnets of telegraphic instruments,

$$M = knC\sqrt{d};$$

M magnetic intensity.
n number of turns of wire.
C strength of the current.
d diameter of the core.
k a constant.

We may here remark that the maximum value for M occurs when n has a certain value such that the resistance of the coil of the electro-magnet is equal to the resistance of the rest of the circuit. This is in accordance with the doctrine of the conservation of energy, as it shows, in other words, that the power of the electro-magnet is a maximum when the electric energy expended in its coils is a maximum. This condition enables us to calculate for each particular case the value of C and the resistance R of the wire to be coiled on the electro-magnet.

From this consideration alone we should be led to indefinitely increase the dimensions of the core and of the wire, since for any given e. m. f. M increases with d and n if R be kept constant, so that C may remain the same. In practice, the indeterminate nature of the problem as thus expressed, is overcome by introducing another factor, for example, the weight of copper wire to be coiled on the magnet, or, more simply, the space to be occupied by the wire on the reel. Let V be this space, determined by a cylindrical annular space of known dimensions.

Let s be the sectional area of the wire, l its length, a its specific resistance.

The condition that the resistance of the coils must be equal to R, gives

$$R = \frac{la}{s},$$

and the space occupied by the wire (V being affected by a practical co-efficient, taking into account the thickness of the covering of the wire and the space between the turns) gives

$$V = ls.$$

From these two equations may be deducted

$$l = \sqrt{\frac{VR}{a}};$$

$$s = \frac{V}{l};$$

Knowing the sectional area s of the wire, its diameter d is obtained by the formula

$$d = \sqrt{\frac{4s}{\pi}}.$$

Practically these formulæ enable the length and diameter of the wire required to fill the reel of an electro-magnet, of which the dimensions are known, so that the coil may have a given resistance, to be calculated. This is the most general case.

We know of no simple practical formulæ which enable the dimensions of the different parts of an electro-magnet to be calculated so that it may carry a given weight, or produce a magnetic field of given intensity.

ELECTRO-MAGNETS OF TELEGRAPH INSTRUMENTS.

The problem in this case is simplified, because the dimensions of the reel are given, and are determined by considerations which are independent of the absolute strength of the magnetic field to be produced. The indeterminate nature of the general problem thus disappears.

Let

A be the area of the section of the wire space on the reel on one side only of the axis made by a plane passing through the axis, and at right angles to the turns of wire.

l the mean length of a turn of wire in millimètres.
ρ radius of the wire in millimètres.
ϵ thickness of the covering of the wire in millimètres.
c conductivity of the wire (pure copper $= 1$).
r resistance of the coil in ohms.
t total number of turns.
a a coefficient depending on the mode of winding.

If the layers are exactly superposed, A is supposed to be divided into squares, and $a = 4$; if the turns of one layer lie in the interstices of the one underneath, A is divided into hexagons, and $a = 2\sqrt{3} = 3·46$. If the layers are separated by some insulating material (paraffined paper or thin guttapercha cloth), ϵ is increased by one quarter of the thickness of this material.

Then
$$= \frac{A}{a(\rho + \epsilon)^2},$$
$$r = \frac{tl}{\pi c \rho^2} = \frac{Al}{\pi d c \rho^2 (\rho + \epsilon)^2}.$$

To express r in ohms a coefficient must be introduced, and the formula becomes
$$r = \frac{lt}{194456\rho^2} \times \frac{1}{c}.$$

To find the resistance of a coil of wire of given thickness wound on a reel A.—Say the coil of the electro-magnet of a Morse receiver: the reel being 60 millimètres long, wire space 10 mm. deep, and the core 10 mm. in diameter, wound with wire No. 32 of the *jauge carcasse;* for which
$$2\rho = ·16 \quad 2(\rho + \epsilon) = ·20.$$
∴ thickness of the silk $\epsilon = ·02$ mm.

Suppose it to be so wound that $a = 4$,
$$A = 60 \times 10 = 600; \quad l = 20\pi.$$

Then for the number of turns t,
$$t = \frac{A}{a(\rho + \epsilon)^2} = \frac{600}{4 \times ·01} = 15{,}000.$$

and for the resistance r,
$$r = \frac{lt}{194456\rho^2} \times \frac{1}{c} = \frac{600 \times 15000}{194456 \times 0·8^2} \times \frac{1}{c} = 759 \times \frac{1}{c} \ ohms.$$

If $c = 1$ $\qquad r = 759\ ohms,$

if $c = ·9$, about the usual practical value for commercial wire,
$$r = \frac{759}{·9} = 843\ ohms.$$

When the dimensions of the reel, the number of turns of wire t, the resistance r of the coil, and the diameter 2ρ of the bare wire are known, the conductivity c of the wire, and the thickness ϵ of the covering can be calculated.

To calculate the diameter 2ρ of a wire of conductivity c, with a covering of thickness ϵ, the resistance of the coil being r ohms.
Let

$$b = \sqrt[4]{\frac{At}{\pi a c r}}.$$

The value of b is calculated, then,

$$\rho = \sqrt{b^2 + \frac{\epsilon^2}{4}} = \frac{\epsilon}{2}.$$

But ϵ being very small $\frac{\epsilon}{4}$ may be neglected; then,

$$\rho = b - \frac{\epsilon}{2} \text{ or } 2\rho = 2b - \epsilon.$$

Thickness of wire with which a galvanometer or electro-magnet coil must be wound so as to obtain the maximum magnetic effect with a given external resistance.

It can be shown by calculation that if 2ρ be the diameter of the bare wire, $2(\rho + \epsilon)$ the diameter of the wire with its covering, r the resistance of the coil, and R the external resistance, that the best effect is produced when

$$2\rho : (\rho + \epsilon) :: r : R,$$

$$\text{or } \frac{r}{R} = \frac{\rho}{\rho + \epsilon}.$$

Construction of the reels of electro-magnets. —It is better to make them of box or ebonite than of copper or brass, but if of metal they should be split; it is also as well to cut a groove about 2 mm. wide and 2 mm. deep, in the core, to hasten demagnetisation. During the *winding*, the turns of wire must be well insulated from each other, and metallic filings, which would penetrate the coating, must be carefully excluded. If a disc be fixed in the middle of a reel and the two halves of the wire be wound one on each side of it, starting from the disc, the two ends of the wire will both come out at the surface of the coil, and thus the inconvenience is avoided which is so often felt when the end of the wire leading from the innermost layer of the coil breaks off (*Culley*).

PRODUCTION AND APPLICATIONS OF ELECTRICITY.

The apparatus used for the *production* of electrical energy may be divided into three classes:

1st. *Batteries* which convert *chemical* energy into electrical energy;

2nd. *Thermopiles*, which convert *heat* energy into electrical energy;

3rd. *Dynamo and magneto machines*, which convert *mechanical* energy into electrical energy.

The applications of this energy may always be reduced to the production of chemical, heating, or mechanical effects.

The production of electrical energy is, strictly speaking, only the conversion of one form of energy into another form of energy, one mode of motion into another mode of motion. This reciprocity of cause and effect may be expressed by saying that producers or generators of electricity are *reversible* and *interchangeable*, that is to say, that any one of them can be the seat of analogous but reciprocal actions to those for which it is originally constructed, and that they may be substituted one for the other for the production of the same effects. Thus, a magneto-electric machine when it is put in motion produces a current at the expense of mechanical work, or sets itself in motion and produces mechanical work if a current be supplied to it from some other source of electrical energy. This is expressed by saying that a magneto-electric machine is reversible.

The magneto machine may also be set in motion by the current furnished by a battery or thermopile, the three sources of electricity may be substituted one for the other; that is to say, they are interchangeable. The applications of electricity may be divided into three large groups characterised by the nature of the actions produced.

Chemical actions.—All forms of apparatus which produce electrical energy at the expense of chemical action are called *batteries*. *Accumulators* are *reversible* batteries.

The chemical actions produced by currents are embraced under the general title of *electrolysis*. The applications of electrolysis are electro-metallurgy, including electrotyping, electro-gilding, silvering, nickel plating, etc., and a form of colouring in metal, a new process which has been used at Nuremberg for the decoration of metallic toys. Electrolysis has also been applied to the purification of alcohol, and the more rapid amalgamation of gold in the quartz-crushing method of obtaining that metal.

Heating effects.—Apparatus which convert heat into electrical energy are called *thermopiles*.

The heating effects of currents have received the following applications. Melting of metals (but little used as yet), firing of fuses (for mines, torpedoes, blasting, etc.), safety catches for electric light, and electric power leads, and arc and incandescent electric lights.

Mechanical actions.—Two different types of machines exist; the first, called *static* machines, produce electrical energy in the form of *charges*, either by friction, or by influence, or *electrostatic induction*. They have no practical application as yet, though the Voss induction machine is reversible; that is to say, that if the two conductors of such a machine be kept charged by means of another it will rotate (*Silvanus Thompson*).

The second type is based on the *magnetic* action of currents, and electro-magnetic *induction*. Magneto and dynamo-electric machines convert mechanical into electrical energy, whilst electrical energy produces mechanical energy, work, or motion in bells, railway signals, clocks, telegraph instruments, telephones, the regulating mechanism of arc lamps, and the motors which are used in the transmission of power to a distance, etc.

Other effects of currents.—The mutual actions of light and electricity form the as yet almost purely scientific branch, electro-optics, but which, however, in the case of the change of resistance of certain bodies under the influence of light, has given rise to various beginnings of practical applications, such as Graham Bell's *photophone*, the *radiophone*, the *teleradiophone*, and the *teleradiophone multiple auto-reversible* of M. Mercadier, etc.

The *physiological action* of currents has lately been much studied, and its effects are of daily use in the physiological laboratory. Some of the physiological effects are now recognised as infallible means of diagnosis in certain diseases, and others have been used with much success, though without any very great certainty, as means of cure. In the case of lead palsy, or painter's dropped wrist, the action of interrupted currents enables cures to be effected in cases which would formerly have been hopeless.

These two branches of the subject are foreign to the nature of this work, and must be studied in special treatises.

In the following pages the order of the subdivisions just laid down will be followed as strictly as possible, the reversible and interchangeable nature of the phenomena, however, make some small departures from this order inevitable.

BATTERIES.

These will be divided into three distinct groups:
One-fluid batteries with no depolariser.
One-fluid batteries with solid or liquid depolariser.
Two-fluid batteries.

ONE-FLUID BATTERIES WITH NO DEPOLARISER.

Volta's battery (1800).—Plate of zinc, plate of copper, sulphuric acid diluted with sixteen times its volume of water. The current is produced by the oxydation of the zinc and its conversion into zinc sulphate. It polarises rapidly when the circuit is closed; the hydrogen sticks to the copper plate, the sulphate of zinc is decomposed and produces a deposit of zinc on the copper. The first form was the Volta's *pile*, so called because it consisted of a pile of plates arranged thus: copper, flannel, zinc, copper, flannel, zinc, and so on. Modified by *Cruikshank* (1801) under the form of the trough battery, then the separate trough battery of *Volta*, who had already employed the same principle in his "crown of cups;" the windlass battery, due to *Crahay* (1841). The *spiral* battery was invented in 1821 by *Offershaus*, with the view of diminishing the internal resistance of the elements; *Wollaston* (1815) increased the surface of the copper, leaving the zinc unchanged. *Münch* (1841) constructed a trough battery with no partitions between the elements, but short circuits occur between the elements; *Faraday* (1835) prevented this by separating the elements by sheets of well-varnished paper. *Pulvermacher* (1857) constructed *galvanic chains* formed of zinc and copper wires coiled on cylinders of porous wood, which absorb the exciting fluid (vinegar and water) when the chain is dipped into it.

Amalgamation of the zincs prevents the formation of local couples due to impurities in the zinc. This fact was discovered by *Kemp* (1828), and applied to batteries by *Sturgeon* (1830).

Modifications of the exciting fluid.—In order to prevent polarisation, many exciting fluids have been proposed; sulphate of copper, chromic acid, oxygenated water, sal-ammoniac, etc.

Modifications of the copper plate in order to diminish the polarisation, *Poggendorff* (1840).

1st. Heated the copper in air until the colours which appear at first had passed away.

2nd. Dipped the copper in nitric acid and then washed it with water.

3rd. Covered the plate with a powdery deposit of copper by electrolysis.

Page (1852) pierced the plate full of holes and covered it with an electro-deposit of rough copper.

Walker (1852) electro-deposited copper on the plate, allowing the solution to become almost exhausted, or formed the plate of copper wire gauze.

Carbon battery plates.—Retort carbon was used in one-fluid batteries by *de Leuchtenberg* (1845) and *Stöhrer* (1849), a solution of alum being employed. Carbon agglomerates are becoming more used than retort carbon, which, though cheap in itself, is expensive to work.

Smee (1840).—Copper plate replaced by platinised platinum, or better, platinised silver. The e. m. f. = ·47 volt, about. Used for electrotyping.

Walker (1859). — Platinised carbon. Cheaper than Smee's, e. m. f. = ·66 volt.

Maiche (1879).—Fragments of platinised carbon only partly immersed in the fluid. Scraps of zinc in a bath of mercury. Water acidulated with sulphuric acid saturated with common salt or sal-ammoniac. E = 1·25 volts with common salt. Quantity small, high internal resistance.

Iron positive plates.—*Sturgeon* (1840) used cast iron with water acidulated with one-eighth its volume of sulphuric acid: *Münnich* (1849) used amalgamated iron; *Callan* (1855) a flattened cast-iron vessel and slightly diluted pure hydrochloric acid.

ONE-FLUID BATTERIES WITH SOLID DEPOLARISER.

Warren de la Rue (1868).—Unamalgamated zinc, sal-ammoniac, silver surrounded by chloride of silver. E = 1·03 volts. The invention of this battery is also claimed for Marié-Davy.

Skrivanow (1883).—A pocket battery formed of a plate of zinc and chloride of silver wrapped in parchment paper immersed in a solution of 75 parts caustic potash and 100 parts of water. An element weighing 100 grammes has an e. m. f. of 1·45 to 1·5 volts. It can give out 1 ampère for one hour. The potash solution must be renewed after this work, and the chloride of silver must be replaced after the potash has been renewed two or three times.

Gaiffe.—Zinc (not amalgamated); silver, surrounded by chloride of silver; 5 per cent. solution of chloride of zinc. $E = 1·02$ volts. Used as a standard cell with condensers and electrometers.

Marié-Davy.—Amalgamated zinc, acidulated water, carbon, and paste of sulphate of mercury. $E = 1·52$ volts.

Leclanché (1868).—Amalgamated zinc, sal-ammoniac, carbon surrounded by fragments of carbon ("carbon shingle"), mixed with black oxide of manganese (needle form). $E = 1·48$ volts when the battery is not polarised.

POROUS POT ELEMENTS.

	SMALL.	MEDIUM.	LARGE (DISC ELEMENT.)
Diameter of porous pot in centimètres	6	6	8
Height	11	15	15
Internal resistance in ohms (min.)	9 to 10	5 to 6	4
Annual chemical work in grammes of copper	40	60 to 70	100 to 125
Annual chemical work in coulombs	60,000	100,000	150,000

Agglomerate Leclanché batteries.—The depolariser is formed by one or more agglomerate plates, kept in position against the carbon by indiarubber bands. This form is handier, more economical, and of less internal resistance. $E = 1·48$ volts.

No. 1, with one plate . . . . . $r = 1·8$ ohms.
No 2, with two plates . . . . . $r = 1·4$,,
Disc element with three plates . . . . $r = ·9$,,

Agglomerate.—A paste of 40 parts black oxide of manganese, 52 of carbon, 5 of gum lac, and 3 of bisulphate of potassium, compressed by a pressure of 300 atmospheres at a temperature of 100° C.

Lalande and Chaperon's oxide of copper battery (1882).—Zinc; 30 or 40 per cent. solution of caustic potash; binoxide of copper in contact with a plate of iron or copper. E. m. f. ·8 to ·9 volt, according to the external work. Practically inactive on open circuit. The current varies with the dimensions of the element. The small hermetically sealed form can give 40,000 coulombs at the rate of

·1 to ·2 ampère. The spiral form 200,000 coulombs and ·5 ampère. The small trough form 500,000 coulombs and 1·5 ampères. The large trough form 1,000,000 coulombs and 6 to 8 ampères. To prevent the caustic potash from absorbing carbonic acid, the trough elements have the fluid covered by a layer of heavy petroleum.

ONE-FLUID BATTERIES WITH LIQUID DEPOLARISER.

The type of these batteries is *Poggendorff's bichromate of potassium* battery. The shape and dimensions of the battery and the composition of the fluid have been varied by different makers, in order to produce particular effects.

Poggendorff's formula (1842).—100 grammes of bichromate of potash dissolved in one litre of boiling water, with 50 grammes of sulphuric acid added.

Delaurier's formula.—Water, 200 grammes; bichromate of potash, 18·4 grammes; sulphuric acid, 42·8 grammes. This formula is that given by the chemical equivalents:

$$KO2CrO_3 + 7SO_3 + 3Zn = 3ZnSO_3 + KOSO_3 + Cr_2O_33SO_3.$$

The final products are solution of zinc sulphate and a chrome alum.

Chutaux's formula (1868).—The fluid (which is kept circulating through the cells of Chutaux's battery) is composed of

Water	1·00	grammes.
Bichromate of potash	100	,,
Bisulphate of mercury	100	,,
Sulphuric acid at 66°	50	,,

Dronier's salt.—A mixture of one-third bichromate of potash and two-thirds bisulphate of potash; when dissolved in water, this mixture forms the exciting fluid.

Special modifications of Poggendorff's battery. —For short experiments, M. *Grenet* devised the bottle battery. The zinc is dipped into the fluid during the experiment, and then withdrawn. *Trouvé*, *Gaiffe*, and *Ducretet* suspend the plates from a windlass, so that they may be withdrawn from the fluid, or more or less immersed at will.

Trouvé's cell (1875).—One zinc and two carbons, active surface 15 centimètres inside.

During the "spurt" at the beginning, $E = 2$ volts ; $r = \cdot 0016$ ohms.
After the "spurt" . . . . $E = 1\cdot 9$ volts ; $r = \cdot 07$ to $\cdot 08$ ohms.

On short circuit, one element gave 24 ampères for 20 minutes without polarising (*d'Arsonval*). A fully charged element can give 180,000 coulombs ($=$ 50 ampère hours) before the solution becomes exhausted.

Tissandier's cell (1882), giving a very large current, 100 ampères through an external resistance of $\cdot 01$ ohm.

SOLUTION.

Water	100 parts by weight.
Bichromate of potash	16 ,, ,,
Sulphuric acid 66°	37 ,, ,,

The bichromate to be reduced to fine powder, taking care not to inhale the dust, which produces ulceration of the lining membranes of the nose. Part of the bichromate is dissolved in water at about 40° C. in an earthen vessel; the acid is then added, and the mixture is violently stirred until the whole of the bichromate is dissolved. It must be allowed to cool down to 35° C. before being used. Below 15° C. the liquid works badly. This battery gives more than one kgm. of useful electrical energy per kilogramme of weight for two to three hours, and effective mechanical work of more than a horse-power for the same length of time for a weight of about 200 kilogrammes (24 elements in series, and a Siemens dynamo as motor).

TWO-FLUID BATTERIES.

Becquerel (1829).—Zinc, nitrate of zinc; bladder; copper, and nitrate of copper.

Daniell (1836).—Zinc, acidulated water; bladder or porous pot; copper, saturated solution of sulphate of copper. $E = 1\cdot 079$ volts. Action goes on even on open circuit, the solution is maintained by adding sulphate of copper. The solution round the zinc may be pure water, salt and water, or a solution of sulphate of zinc. There are many modifications of the Daniell cell.

Trough battery.—Flattened elements arranged in a trough with partitions. The Post-Office pattern is a teak trough, marine glued inside to make it water-tight, with alternate diaphragms of slate and porous earthenware.

Eisenlohr (1849) replaced the dilute sulphuric acid by a **solution of tartar** (sodium bitartrate).

Minotti's cell.—The porous pot is replaced by a layer of sand or sawdust, the copper and copper sulphate below and the zinc above.

Gravity battery.—The two liquids are kept separate by their difference of density, there is no porous diaphragm. Called *Meidinger's* (1859) in Germany and *Callaud's* (1861) in France. It is known under both names.

Sir W. Thomson's tray battery.—Large-surfaced horizontal elements, the zinc in the form of a grating enveloped in parchment paper so as to form a tray to contain the sulphate of zinc solution; very low internal resistance.

E. Reynier (1880).—Porous cell made of parchment paper so folded as to have no seams.

Amalgamated zinc; solution, 300 parts of caustic soda to 1,000 parts of water; copper, solution of sulphate of copper containing sulphate of soda or sulphuric acid. The conductivity of the solutions is increased by the addition of small quantities of different alkaline salts. E. m. f. $= 1\cdot 5$ volts. Resistance of a 3-litre cell with lined parchment-paper porous cell, $\cdot 075$ ohm.

E. Reynier (1881).—External cell of copper of flat form, forming the positive plate. The bottom is provided with a wooden flooring. Jacketed zinc. Only the sulphate of copper solution requires to be renewed. The sulphate of zinc passes through the jacketing by itself by osmosis.

Grove (1839).—Zinc (amalgamated), dilute sulphuric acid; porous pot; nitric acid, platinum. $E = 1\cdot 96$ volts. The surface of the platinum may be increased by bending it into an S shape (*Poggendorff*, 1849).

Callan (1847) replaced the platinum by platinised lead, and the nitric acid by a mixture of 4 parts concentrated sulphuric acid, 2 of nitric acid, and 2 of a saturated solution of nitrate of potash.

Bunsen (1840) replaced the platinum by artificial carbon in the form of a hollow cylinder. $E = 1\cdot 9$ volts.

Archereau (1842).—Put the block of carbon in the middle and the zinc outside.

Bunsen's cell (*d'Arsonval*).—Amalgamated zinc in dilute sulphuric acid (one-twentieth by volume). In the porous pot a plate of carbon, in common commercial nitric acid, sp. gr. 36° to 40° Baumé.

An element 20 cm. high has an internal resistance of $\cdot 08$ to $\cdot 11$ ohm.

The e. m. f. $= 1\cdot 8$ volts. When the nitric acid is about 30° Baumé the battery runs down rapidly. The Bunsen element consumes 1·3 grammes of zinc per ampère per hour (Faraday's law gives 1·295). Nitric acid of density 36° Baumé contains 45 per cent. of anhydrous nitric acid; it will work until it has fallen to 28° Baumé. Under these conditions only 130 grammes of acid are utilised per kilogramme expended. The weight of **acid expended is at least ten** times that of the zinc consumed.

D'Arsonval's depolarising liquid (to be used instead of nitric acid, and in the same way, 1879).

Nitric acid	1 part.
Hydrochloric acid	1 ,,
Water	2 ,,

is used in circulating batteries, zinc in the porous cell, the positive pole being formed by a crown of carbon rods 1 cm. in diameter. The **current is constant, the internal resistance reduced to a minimum, and the depolarising surface very large.** An element 20 cm. high can give **as** much as 40 ampères on short circuit.

Carbons for Bunsen cells.—Carré's carbons are better than retort carbon; they have a very high conductivity, and their closeness of texture prevents the acids from escaping by capillary attraction, and so attacking the metallic connections. This fault can be completely overcome **by** steeping the upper part of the carbon for several minutes in **melted paraffin.** When cold electroplate it with copper, and then dip it into melted type metal. **Perfect and indestructible contact can thus** be insured (*d'Arsonval*).

D'Arsonval's zinc carbon element differs from **the** Bunsen cell by the composition of the fluids.

Fluid round the zinc, or exciting fluid.

Water	20 volumes.
Sulphuric acid (purified by oil; see page 297)	1 ,,
Common hydrochloric acid	1 ,

Fluid round the carbon or depolariser.

Common nitric acid	1 volume.
Common hydrochloric acid	1 ,
Water acidulated with $\frac{1}{10}$th of sulphuric acid	2

This cell does not polarise on short circuit; its e. m. f. is as much as **2·2 volts.**

BATTERIES.

Couples in which the negative electrode is jacketed (*cloisonné*) (*E. Reynier*).—In most two-fluid combinations the negative electrode is immersed in a solution of some salt of the metal of which it is composed.

If the negative **electrode be jacketed,** that is to say, provided with a tight **covering not attacked by the** liquids in the battery, but yet **permeable by them, a constant** couple may be formed by simply dipping a **positive electrode in** the ordinary state into the depolarising liquid alongside of the jacketed negative electrode.

The small quantity of liquid confined in the jacket against the negative **electrode** is soon charged with a salt of its metal; the couple then **acts like** a two-fluid cell, the excess of salt being eliminated through the **jacket** as fast as it is formed **by a process of osmosis, the rate of which is auto**matically regulated **by the current.**

This plan **reduces the maintenance of constant batteries to the keeping in** good order **of one fluid only; it has been applied with success to the** combination **zinc, sulphate of zinc;** sulphate of copper, copper. The cell being composed **of a** copper **vessel and** a jacketed zinc. (If the sulphate of copper be replaced by "**gilder's** verdigris" the battery becomes the most economical known.)

E. m. f. of some jacketed zinc cells (*E. Reynier*):

	New volts.	Polarised volts.
Common zinc jacketed, bare copper, dilute sulphuric acid	·848	·411
Common zinc jacketed, bare iron, dilute sulphuric acid*	·401	·409
Amalgamated zinc jacketed, bare **iron, dilute** sulphuric acid	·466	·466
Amalgamated zinc jacketed, bare iron, 20 per cent. solution of bisulphate of soda . .	·504	·509

M. Reynier **has tried to** apply jacketing to bichromate cells; **but he** has not been **able** to construct any flexible jacket which will resist **the** destructive action of the bichromate mixture.

Modifications of Grove's and Bunsen's cells.—Efforts have been made, **but as** yet without success, to find a substitute **for zinc** in these cells. *M. Rousse* proposed lead or iron attacked by **nitric** acid. *M. Maiche* (1864) a cylinder of sheet iron attacked by water containing one-hundredth of nitric acid, etc. Attempts have also been

---

* *Observation.*—The **couple zinc, acidulated water, iron, is perfectly constant.**

made to replace the platinum by passive iron (*Hawkins*, 1840; *Schönbein*, 1842). Concentrated nitric acid or aqua regia is necessary for this purpose.

Attempts have also been made to replace the *depolariser* (nitric acid) by other bodies: chromate of potassium, *Bunsen* (1843); chlorate of potash, *Leeson* (1843); chromic acid, chloric acid, perchloride of iron, picric acid, etc.

Marié-Davy's cell.—A Bunsen cell in which the nitric acid is replaced by a paste of sulphate of mercury protoxide (Hg_2O, SO_3) packed round the carbon. $E = 1·2$ volts. Answers for intermittent work, requires very porous cells, weak solutions of high resistance.

Duchemin (1865).—Zinc in a solution of common salt, the nitric acid replaced by a solution of perchloride of iron. The solution is refreshed by passing a stream of chlorine through it. $E = 1·54$.

Delaurier (1870).—Nitric acid replaced by:

Chromic acid (4 equivalents)	25·14 parts.
Sulphate of protoxide of iron (1 equivalent)	25· ,,
English sulphuric acid	30·62 ,,
Water	60· ,,

The hydrogen is absorbed, and a mixture of sulphate of protoxide of iron, and sulphate of sesquioxide of chrome is formed.

Bichromate of potash.—Depolarising liquid of 100 parts of water, 25 of sulphuric acid, and 12 of bichromate; porous pot, amalgamated zinc, water acidulated with one-twelfth its weight of sulphuric acid. $E = 2·03$ volts for the first few seconds. In *Fuller's* battery the central zinc stands in a small pool of mercury to keep up the amalgamation. Water only is in practice put into the porous pot with the zinc. $E = 2$ volts.

In the element invented by *Cloris Baudet* (1879), there is a supply of sulphuric acid and bichromate arranged in the outer pot with the carbon, so as to keep up the strength of the depolarising liquid. $E = 2$ volts: $r = ·22$ to $·3$ ohm for the 20 cm. form.

In *Higgins'* cell the zinc, which is perfectly amalgamated, is immersed in a solution of sulphuric acid ($\frac{1}{20}$th by volume), the carbon plate in a chromic solution containing 45 parts water, 15 sulphuric acid, and 5 of bichromate of potash (by weight). $E = 2·2$ volts. The 15 cm. high element has an internal resistance of ·4 to ·5 ohm.

D'Arsonval's bichromate cell.—Porous pot full of broken-up retort carbon, plain water with the zinc; the depolarising liquid is:

Water saturated in the cold with bichromate of potash .	1 volume.
Common hydrochloric acid	1 ,,

This liquid should flow continuously through the cell; this element makes no smell, and is always ready when wanted.

Niaudet's chloride of lime battery (1879).—Plate of zinc in 24 per cent. solution of common salt; plate of carbon in porous pot, with fragments of carbon and chloride of lime (bleaching powder), the depolarising agent is the hydrochlorous acid. Initial e. m. f. $=$ 1·65 volts after being left alone for several months. $E = 1·5$ volts, r of the common form $= 5$ ohms. Action only takes place when the circuit is closed, but smells disagreeably, so that the cells have to be hermetically sealed.

Circulation, agitation, and blowing air through the cells are three excellent ways of disengaging hydrogen from the carbon plates, and bringing oxygen in contact with them. Agitation has been employed by *Chutaux*, *Camacho*, etc.; blowing air through by *Grenet* and *Byrne*.

Thermo-chemical batteries.—The current is produced by the oxydation of carbon at a high temperature under the action of nitrate of potassium or sodium. The fundamental experiment is due to *Becquerel* (1855), repeated by *Jablochkoff* (1877), and taken up again by *Dr. Brard* (1882), who has produced an electrogenic torch which produces a current as it burns. This new generator of electricity has not yet come into use, we therefore only notice it.

Electromotive forces of one-fluid batteries with no depolariser (reduced to volts from observations by *Poggendorff* and *E. Becquerel*):

	Open circuit.	After polarisation.
Copper, dilute sulphuric acid, common zinc .	·81	—
Silver, ,, ,, ,, ,, ,, .	1·03	—
Copper, ,, ,, ,, amalgamated zinc	·94	·44
Silver, ,, ,, ,, ,, ,, .	1·24	·52
Platinum, ,, ,, ,, ,, ,, .	1·44	·65

ELECTROMOTIVE FORCE OF GROVE'S CELL (*Poggendorff*).

Fluid round the Zinc.	Fluid round the Platinum.	Electromotive Force in Volts.
Sulphuric acid:		
Density . . = 1·136	Fuming nitric acid . .	1·955
,, . . = 1·136	Nitric acid, d . = 1·33	1·809
,, . . = 1·06	,, ,, . = 1·33	1·730
,, . . = 1·136	,, ,, . = 1·19	1·681
,, . . = 1·06	,, ,, . = 1·19	1·631
Solution of sulphate of zinc	,, ,, . = 1·33	1·673
,, of common salt .	,, ,, . = 1·33	1·905

ELECTROMOTIVE FORCES OF AMALGAMS OF POTASSIUM AND ZINC.
(The amalgams were enclosed in porous cells) (*Wheatstone*).

Amalgam.	Solution.	Positive Pole.	Electromotive Force.
Potassium	Sulphate of zinc.	Zinc.	1·043
	Sulphate of copper.	Copper.	1·122
	Chloride of platinum.	Platinum.	2·482
	Sulphuric acid.	Peroxide of lead.	3·525
	Sulphuric acid.	Peroxide of manganese.	2·921
Zinc	Sulphate of copper.	Copper.	1·079
	Nitrate of copper.	Copper.	1·043
	Chloride of platinum.	Platinum.	1·438
	Sulphuric acid.	Peroxide of lead.	2·446
	Sulphuric acid.	Peroxide of manganese.	1·942

Electromotive forces of cells containing only one electrolyte (*E. Reynier*).—The electromotive force of such cells is very variable; it decreases when the circuit is closed, and increases when the battery is at rest; for the same voltaic combination, it appears to be higher when the surface of the positive plate is very large compared to that of the negative plate. Thus the *apparent* e. m. f. varies with the design of the cell, the circumstances of the experiment, and the method of measurement employed. The two values of the e. m. f. which must be known are the highest and the lowest. M. Reynier has measured these two extreme values by means of two different types of cell, specially constructed for this purpose, which he calls *pile à maxima* and *pile à minima*. The positive plate of the first type has 300 times more surface than the negative, whilst in the second type the negative plate exposes far more surface than the positive. The e. m. f. is measured after the cell has been for some long time on short circuit. The following are the figures which he has obtained for some voltaic combinations.

LIQUID.	DESCRIPTION OF CELL.		ELECTROMOTIVE FORCE IN VOLTS.	
	NEGATIVE PLATE.	POSITIVE PLATE.	MAXIMUM.	MINIMUM.
Water acidulated with sulphuric acid. Water, 1000 vols. Monohydrated sulphuric acid, 2 vols.	Common zinc.	Platinum.	—	·5
	Amalgamated zinc.	Platinum.	—	·561
	Common zinc.	Silver.	—	< ·098
	Amalgamated zinc.	Silver.	—	·108
	Common zinc.	Carbon.	1·22	·04
	Amalgamated zinc.	Carbon.	1·26	·226
	Common zinc.	Lead.	·55	·144
	Amalgamated zinc.	Lead.	·684	·152
	Common zinc.	Copper.	·94	·194
	Amalgamated zinc.	Copper.	1·072	·272
	Common zinc.	Iron.	·429	·309
	Amalgamated zinc.	Iron.	·476	·323
	Amalgamated zinc.	Common zinc.	—	·09
	Iron.	Copper.	·5	—
Solution of sodium chloride. Water, 1000 gr. Sodium chloride, 250 gr.	Common zinc.	Platinum.	—	·034
	Common zinc.	Carbon.	1·08	< ·04
	Common zinc.	Silver.	—	·043
	Common zinc.	Copper.	·78	·025
	Amalgamated zinc.	Copper.	·82	—
	Common zinc.	Iron.	·378	·046
	Amalgamated zinc.	Iron.	·469	—
	Common zinc.	Lead.	·503	·044
	Amalgamated zinc.	Lead.	·52	—
	Iron.	Copper.	·26	—
	Lead.	Copper.	·26	—
Zinc Chloride. Water, 1,000 gr. Zinc chloride, 110 gr.	Common zinc.	Copper.	·85	—
	Amalgamated zinc.	Copper.	·86	—
Zinc sulphate. Water, 1,000 gr. Zinc sulphate, 500 gr.	Common zinc.	Copper.	·998	—
	Amalgamated zinc.	Copper.	1·04	—
Caustic soda. Water, 1,000 gr. Caustic soda, 250 gr.	Common zinc.	Copper.	1·06	—
	Amalgamated zinc.	Copper.	1·09	—

Electromotive forces of some two-fluid cells
(*E. Reynier*).

	Volts.
Standard *Daniell*.—Unamalgamated zinc, sulphate of zinc $d = 1\cdot09$. Copper, sulphate of copper $d = 1\cdot16$	1·068
The same, a very small quantity of sulphuric acid added to the sulphate of copper	·993
The same, with a very small quantity of sulphuric acid added to both liquids	·929
The same, with a small quantity of tartaric acid added to the sulphate of copper	1·015
Unamalgamated zinc, caustic potash, 30 per cent. solution. Copper, sulphate of copper $d = 1\cdot16$	1·555
Unamalgamated zinc, solution of soda and potash (potash 175, soda 250, water 1000). Copper, sulphate of copper $d = 1\cdot16$	1·661
Unamalgamated zinc, solution of soda (Reynier's formula). Copper, solution of sulphate of copper (Reynier's formula)	1·473
Amalgamated zinc, solution of soda (Reynier's formula). Copper, solution of sulphate of copper (Reynier's formula)	1·5
Iron, commercial sulphate of iron $d = 1\cdot20$. Copper, sulphate of copper $d = 1\cdot16$	·711

Theoretical electromotive forces.—Calculated from the electro-chemical equivalents.

Smee's cell	·886 volt.
Daniell's cell	1·156 ,,
Grove's cell	1·991 ,,

Direct observation always gives lower values than these, because of the secondary actions which are always going on in the cells.

Theoretical conditions of a perfect battery.—
1. High e. m. f.
2. Small and constant internal resistance.
3. Constant e. m. f., no matter how large the current.
4. Consumption of cheap substances.
5. Chemical action always proportional to the output of energy, and consequently no consumption when the circuit is open.
6. Convenient and practical form enabling the state of the battery to be easily observed, and the cells to be easily refreshed when necessary.

No battery fulfils all these conditions. In every case a form of element must be chosen, having qualities suited to the purpose for which it is wanted.

CONSTANTS AND WORK OF SOME KNOWN FORMS OF BATTERY *
(E. Reynier).

CELLS.	E.	r.	WORK PER SECOND.	
			Kgm.	Calories (g.-d.).
Bunsen round, ·2 mètre high	1·9	·24	·384	·888
Rhumkorff, ·2 mètre high	1·9	·06	1·536	3·555
W. Thomson	1·06	·12	·238	·551
Reynier, rectangular form, 3 litres capacity, porous cell of two thicknesses of parchment paper	1·5	·075	·765	1·77
Tommasi, zinc, sulphuric acid, and water ($\frac{1}{20}$), carbon, nitric mixture	1·77	·2	·399	·924

Defects of batteries. — When a battery does not give the expected results, one of the following defects is to be looked for: (1) Exhausted solutions; for example, in a Daniell battery, the sulphate of copper worked out, leaving the solution colourless, or nearly so; (2) bad contacts between the electrodes and the wires, oxydised or badly screwed-up binding screws, etc.; (3) empty or partly empty cells; (4) filaments of metallic deposit causing short circuiting between the battery plates; (5) creeping or deposits of salts forming short circuits either between the plates or from cell to cell. Shaking the cells increases their e. m. f. temporarily by disengaging the gases adhering to the plates. Floating filaments and broken plates give rise to false contacts, which cause the current given by a battery to vary suddenly when it is shaken (*Fleeming Jenkin*).

Choice of a battery for different purposes.—The following list may serve as a guide in most ordinary cases.

Electro-chemical deposition.—Daniell, Smee, Grove, Bunsen, bichromate, Slater.

Gilding.—Daniell, Smee.

Silvering.—Daniell, Smee, Grove, Bunsen, Slater.

Electric light.—Grove, Bunsen, bichromate (Grenet, Cloris Baudet), Tommasi, Carré, Reynier, Accumulators.

---

* The constants were measured on new cells, and the work deduced from them by calculation; as soon as the circuit is closed, the elements vary from these figures in a manner unfavourable to the battery. This is the case even with batteries which are supposed to be constant.

Induction coils.—One-fluid bichromate, Grove, Bunsen; for small pocket coils, sulphate of mercury.

Lecture-room and laboratory experiments.—Bichromate, bottle or windlass form. Grove or Bunsen (in a well-arranged stink chamber), well cared for gravity Daniell.

Medical batteries.—Smee, Trouvé, Onimus, Seure, Leclanché, bichloride of mercury, chloride of silver.

Large telegraph lines.—Daniell, Callaud, Meidinger, Fuller, Leclanché.

Bells and domestic purposes.—Leclanché, sulphate of mercury, sulphate of lead.

Torpedoes.—Leclanché (Silvertown firing battery), special form of bichromate.

Electrical measurements.—Leclanché, bichromate, Daniell's standard.

Standards.—*See* Measuring Instruments (page 62). Practically, only the standard Daniell and Clark's standard cell are used.

CARE AND MAINTENANCE OF BATTERIES.

In the first place, in setting up batteries, all parts should be as nearly as possible chemically clean. The chemicals and the water should be pure. For laboratory work these conditions should be strictly fulfilled, the water used being distilled, and for practical work they should be approached as nearly as possible.

Groves' and Bunsen's cells should be emptied after use, all parts well washed, and kept separately in large pans of clean water, zincs in one, platinums or carbons in another, and porous cells in another. All cells of the Daniell type require careful watching from time to time. A little of the fluid surrounding the zinc should be drawn off with a syringe, and fresh water added. If there be any considerable deposit of copper on the zinc, it must be carefully scraped off, and the solution round the zinc poured away, and fresh added. In practice, water alone is always used for the zinc fluid. The zinc of batteries of the Daniell type should never be amalgamated, as, if copper be deposited on the zinc, it spreads in a film of amalgam over the whole surface.

Gravity cells require some skill in their management. If the blue colour due to the sulphate of copper spreads too high up, a syringe or pipette must be plunged well into the blue solution, and some of it drawn off. Water must then be carefully added on the top, taking care not to agitate the fluid. A funnel with a tube turned up at the end is useful. The tube should be immersed until the turned up end is just below the surface of the fluid, and the water then gently poured into the funnel. A screen of perforated zinc may be suspended a little below the zinc plate to reduce any sulphate of copper which rises too high. From time to time some of the clear fluid near the zinc should be drawn off, and replaced with water, because if the sulphate of zinc solution becomes too strong, the salt is electrolysed, and zinc is deposited on the copper.

All forms of Daniell are liable to creeping of the salts. All deposits of crystals should be cleared away, and all parts of the battery above the fluid should be smeared from time to time with paraffin, wax, vaseline, or hard tallow, which checks the creeping.

A good working rule to prevent exhaustion of the copper solution in all forms of Daniell's cell is to take care that there is always some sulphate of copper undissolved. In gravity cells fresh crystals may be added without disturbing the fluid much, by dropping them down a glass tube of large bore plunged nearly to the bottom of the cell. Leclanché cells require but little attention. Sometimes they creep; the remedy is the same as for the Daniell. They should be examined from time to time, and fresh water added if the quantity of fluid has diminished from evaporation. Care should be taken that there is always some undissolved sal-ammoniac at the bottom of the cell. If the cell be full and there be plenty of sal-ammoniac, and yet it will not work well, it may be that the solution is too rich in zinc chloride. Empty the cell, and put fresh sal-ammoniac and water. If it then fails, the manganese is worked out, and a new cell must be substituted.

Fuller's batteries require some of the zinc solution to be removed and replaced by water from time to time, and occasionally the bichromate solution must be renewed. Take care in emptying the cells not to lose the mercury.

All cells require the zincs to be renewed from time to time. In Daniell's cells the copper deposit is peeled off the copper plates from time to time and preserved, as it commands a good price in the market.

Battery testing.—This is generally roughly done with the linesman's detector, a form of upright galvanometer, wound with one coil of low resistance, called the quantity coil, and one of high resistance, called the tension coil. A linesman soon knows by experience that so many cells of a given type should give so many degrees "quantity" and so many "tension;" *i.e.* he has always at hand a rough index of the internal resistance and electromotive force of the batteries under his charge, and in the event of weakness knows in which direction to work in order to set matters right. The telegraph engineer can, if he wishes, easily calibrate the detectors of his linesmen, so that he can translate their reports into electrical units.

The method used in the Post-office service is of considerable accuracy, and yet easy to carry out. Two resistance boxes are used, one of 10, 25, 50, 100, 200, and 400 ohms; the other of 1·07 (A), 3·21 (B), 4·28 (C), 8·56 (D), 17·12 (E), and 34·24 (F) ohms; *i.e.* in the proportion of 1 : 3 : 4 : 8 : 16 : 32. Also a tangent galvanometer of resistance 1·07 ohms.

Electromotive force test.—First the standard cell (e. m. f. 1·07 volts) is joined direct to the galvanometer, and the deflection brought to about 25°, or a very little under, by means of the directing magnet. The battery to be tested is then put in circuit with the galvanometer and the resistance box of 1·07, 3·21 etc., ohms; if the battery to be tested is a

Number of Cells to be Tested.			Coils to be placed in Circuit in R_1.
Daniells.	Bichromates.	Leclanchés.	
5	—	3	A
10	5	6	B
—	—	8	C
15	—	10	A + C
20	10	12	B + C
25	—	16	A + D
30	15	18 and 20.	B + D
35	—	—	A + C + D
40	20	24	B + C + D
45	—	30	A + E
50	25	32	B + E
55	—	—	A + C + E
60	30	36	B + C + E
—	—	40	D + E
65	—	—	A + D + E
70	35	—	B + D + E
75	—	48 and 50.	A + C + D + E
80	40	—	B + C + D + E
85	—	—	A + F
90	45	60	B + F
95	—	—	A + C + F
100	50	—	B + C + F
105	—	—	A + D + F
110	55	—	B + D + F
115	—	—	A + C + D + F
120	60	—	B + C + D + F
125	—	—	A + E + F
130	65	—	B + E + F
135	—	—	A + C + E + F
140	70	—	B + C + E + F
145	—	—	A + D + E + F
150	75	—	B + D + E + F
155	—	—	A + C + D + E + F
160	80	—	B + C + D + E + F

Daniell, then as many times 1·07 ohms are inserted as there are cells in the battery (taking into account the 1·07 ohms in the galvanometer for one cell), if the battery is in good order the deflection remains the same, if not the e. m. f. can be calculated. For other types of cell, tables are issued showing the most convenient resistances to employ in order to make the calculation easy, and the deflections which ought to be obtained if the cells are in good order.

Test for internal resistance.—After the deflection for electromotive force has been noted, the second resistance box is put as a shunt to the battery; let R_2 be the resistance employed. This reduces the deflection. Let c be the current passing through the galvanometer, then if E be the electromotive force and x the internal resistance of the battery,

$$c = \frac{E}{x + \frac{R_2(R_1 + G)}{R_2 + R_1 + G}} \times \frac{R_2}{R_2 + R_1 + G}$$

$$= \frac{ER_2}{x(R_2 + R_1 + G) + R_2(R_1 + G)},$$

where R_1 is the resistance employed in taking the e. m. f.

The current C passing through the galvanometer previous to the insertion of the shunt is

$$C = \frac{E}{x + R_1 + G},$$

therefore

$$\frac{c}{C} = \frac{(x + R_1 + G)R_2}{x(R_2 + R_1 + G) + R_2(R_1 + G)}$$

$$= \frac{\left(\frac{x}{R_1 + G} + 1\right)R_2}{x\left(\frac{R_2}{R_1 + G} + 1\right) + R_2}.$$

But since x and R_2 are very small compared with $R_1 + G$, we may consider

$$\frac{x}{R_1 + G} \quad \text{and} \quad \frac{R_2}{R_1 + G} \quad \text{equal to 0,}$$

in which case

$$\frac{c}{C} = \frac{R_2}{x + R_2}.$$

If the first deflection (on the degrees scale) be called D and the reduced deflection be called d, then

$$\frac{\tan d}{\tan D} = \frac{c}{C},$$

therefore

$$\frac{\tan d}{\tan D} = \frac{R_2}{x + R_2};$$

or

$$\tan d \times x + \tan d \times R_2 = \tan D \times R_2,$$

therefore

$$\tan d \times x = R_2 (\tan D - \tan d);$$

that is

$$x = R_2 \left(\frac{\tan D}{\tan d} - 1\right).$$

ACCUMULATORS OR SECONDARY BATTERIES.

Though many forms of battery are theoretically reversible, only salts of lead as yet have been found to answer in practice. *M. Gaston Planté* was the first to use lead for a reversible regenerative or secondary battery in 1860. Since these cells have come into practical use they have been called Accumulators, a name which is very unfortunate, being based on an erroneous theory of their action, which, as far as we know, has never been held by any electrician. It has now passed into the language, and must therefore be accepted.

M. Gaston Planté's secondary cell or accumulator (1860).— Two plates of lead, separated by flannel or other absorbent non-conductor, and rolled up in a spiral immersed in a 10 per cent. (by volume) solution of sulphuric acid.

Formation.—A preliminary operation, the object of which is to form as thick a coating of peroxide as possible on the positive plate and to convert as great a depth as possible of the negative plate into spongy or crystalline lead. This is effected by changing the direction of the charging current after discharge, and keeping the cells charged for longer and longer time. In order to shorten the forming process, M. Planté has lately suggested heating the forming bath, or, more simply, immersing the plates for from 24 to 48 hours in nitric acid diluted with half its volume of water. The acid attacks the lead, and makes it more or less porous, thus assisting the action afterwards set up by the charging current.

Electromotive force.—During the first few moments after the stoppage of the charging current, the e. m. f. $= 2\cdot53$ volts. In two minutes it falls to $2\cdot1$ volts, and for two-thirds of the discharge it remains steady at $2\cdot02$ volts. M. Planté explains the excess of e. m. f. at first by the formation of liquid or gaseous compounds rich in oxygen and hydrogen round the electrodes, which tend to decompose or escape very quickly. Their action is added to the normal action, which remains sensibly constant for two-thirds of the discharge.

Internal resistance.—A couple exposing 50 square decimètres of total surface, the plates being five millimètres apart, has an internal resistance of from $\cdot04$ to $\cdot06$ ohm, according to the extent of the formation of the plates.

Total quantity of electricity stored.—A well-formed couple containing 15,000 grammes of lead will deposit 18 grammes of copper in a sulphate of copper voltmeter by its whole discharge. This corresponds to 54,540 coulombs, or 36,360 coulombs per kilogramme of lead. The cell gives out during its discharge from 89 to 90 per cent. of the *quantity of electricity* which has passed through it during its charge, if the discharge be made immediately after the charge. In recent experiments, M. Planté has succeeded in depositing as much as 19 grammes of copper per kilogramme of lead.

Useful discharge.—About two-thirds of the discharge may be utilised without the e. m. f. falling below 2 volts, *i.e.* about 24,240 coulombs. The total energy given out is 4,850 kilogrammètres, or 3,230 kgm. per kg. of lead.

Many modifications of M. Planté's accumulator have been devised to increase the active surface without increasing the weight. (The corrugated plates of MM. *de Kabath, de Pezzer, de Méritens,* etc.)

Faure's accumulator (1881).—Flat or spiral plates of lead covered with minium (red lead) kept in place by parchment paper and felt. The experiments made in January, 1882, by the commission of the Paris Electrical Exhibition, gave the following results:

Thirty-five elements, each weighing $43\cdot7$ kg., electrodes coated with minium 10 kgs. per square mètre. Liquid: water acidulated with one-tenth of its weight of sulphuric acid. The accumulators arranged in series were charged for twenty-four hours forty-five minutes, by a current varying between 11 and $6\cdot36$ ampères, and a mean difference of potential of 91 volts, and received 694,500 coulombs. The charging was effected by a shunt Siemens' dynamo.

The work supplied is thus divided:

Effective work in charging	6,382,100 kgms.
Excitation of field magnets	1,383,600 ,,
Heating of the ring	269,800 ,,
Heating of passive resistances	1,034,500 ,,
Total work given out	9,570,000 ,,

The discharge occupied ten hours thirty-nine minutes, with a mean current of 16·2 ampères, and 61·5 volts difference of potential at the terminals of 12 Maxim lamps in parallel arc. 619,600 coulombs were recovered, the loss was 74,900 coulombs, or about 10 per cent. The external or useful electrical work was 3,809,000 kgm., or 40 per cent. of the total work, and 60 per cent. of the stored work.

The patterns made in England vary from 28 to 45 elements to the ton.

The first require 32 ampères for twelve hours to charge them, the second 20 ampères for twelve hours. Same current given by the discharge, but it may be doubled by halving the time of discharge. For special efforts, then, this current may be largely exceeded for a few moments.

Faure-Sellon-Volkmar accumulators (1882) without felt, etc.; lead plates pierced with holes, or cast-lead gratings; minium, reduced lead, or some lead salt is compressed into the openings. The reduced plates last indefinitely; the oxydised plates are eaten away after about a year's service. A cell containing forty-three plates and weighing 140 kgs., can give 120 ampères for six hours. Another form containing fifty-three plates, and weighing 170 kgs., will melt a copper wire 5 mm. in diameter, and 30 cm. long, which implies a current between 400 and 500 ampères. The good results obtained from the cells manufactured by the Power Storage Company (Limited) are probably as much due to the selection of materials (which is as yet a trade secret), and the care bestowed on their manufacture, as to the fundamental design of the plates.

Copper accumulators (*E. Reynier*).—Positive: peroxydised lead; negative: copper-plated lead; liquid acid solution of copper sulphate: $E = 1.68$ volts. Mr. Sutton, of Australia, laid before the Royal Society a form of copper accumulator: copper; copper sulphate solution; amalgamated lead. It is found that amalgamation makes the peroxydising process more rapid. Professor McLeod, of Cooper's Hill, has observed that the first action of the current is to remove the mercury from the lead in the form of sulphate, the lead surface then begins to peroxydise rapidly, being probably left in a more or less spongy condition. The advantage claimed for copper accumulators is that the colour of the solution is a gauge of the quantity of charge in the cell, being colourless when the cell is fully charged, and deep blue when it is nearly discharged.

Zinc accumulators (*E. Reynier*).—Positive: peroxydised lead; negative: zinc-coated lead; liquid acid solution of zinc sulphate: $E = 2 \cdot 3$ volts. Zinc plates may be used; the difficulty of construction is the necessity for keeping the zinc well amalgamated.

Storing power and output of accumulators.— In order to get good *efficiency* from accumulators their output ought not to exceed half an ampère per kilogramme of total weight; when, however, a rapid output is required, 5 or 6 ampères per kilogramme may be taken at the expense, however, of efficiency. From this point of view the small Planté cells with thin lead plates are the most powerful. The storing power, on the other hand, increases with the dimensions of the cell, and varies in practice between 2,000 and 4,000 kgm. of available electrical energy per kilogramme, or from 70 to 150 kgs. of accumulators per hour horse-power. Theory shows that this weight might be much reduced, but we know of no authentic experiments giving higher results than those quoted here. To avoid heating the accumulators, and so wasting work, the charging current ought not to exceed half an ampère per kg. of accumulators for large sizes, and 2 to 3 ampères per kg. for small. The duration of charging in seconds is given by the ratio of the storing capacity of one element in coulombs to the charging current in ampères. Under the same conditions the duration of the discharge is proportional to the weight of the element.

CALCULATION OF ELECTRO-CHEMICAL DEPOSITS.

When 1 *coulomb* of electricity passes through a decomposition cell it liberates

$\cdot 0105$ milligrammes of hydrogen*.

---

* *Kohlrausch* found by experiment ·010521.
  Mascart „   „ ·010415.

CHEMICAL AND ELECTRO-CHEMICAL EQUIVALENTS.

ELEMENTS.	ATOMIC WEIGHT.	CHEMICAL EQUIVALENT e	ELECTRO-CHEMICAL EQUIVALENT IN MILLIGRAMMES PER COULOMB. z	NUMBER OF COULOMBS NECESSARY TO LIBERATE ONE GRAMME.	WEIGHT IN GRAMMES LIBERATED BY ONE AMPÈRE HOUR.
Electro-positive elements.					
Hydrogen	1	1	·0105	96,000	·0378
Potassium	39·1	39·1	·4105	2,455	1·468
Sodium	23	23	·2415	4,174	·8694
Gold	196·6	65·5	·6875	1,466	2·475
Silver	108	108	1·134	889	4·0824
Copper (cupric)	63	31·5	·3307	3,079	1·19
" (cuprous)	63	63	·6615	1,540	2·38
Mercury (mercuric)	200	100	1·05	960	3·78
" (mercurous)	200	200	2·1	480	7·56
Tin (stannic)	118	29·5	·3097	3,254	1·1149
" (stannous)	118	59	·6195	1,627	2·2298
Iron (ferric)	56	14	·147	6,887	·5292
" (ferrous)	56	28	·294	3,429	1·0584
Nickel	59	29·5	·3097	3,254	1·1149
Zinc	65	32·5	·3412	2,953	1·2283
Lead	207	103·5	1·0867	928	3·9041
Electro-negative elements.					
Oxygen	16	8	·084		
Chlorine	35·5	35·5	·3727		
Iodine	127	127	1·3335		
Bromine	80	80	·84		
Nitrogen	14	4·3	·049		

CHEMICAL EQUIVALENTS.

If e be the chemical equivalent* of an element (hydrogen $= 1$), the weight z of this element set free by one coulomb of electricity will be:

$$z = \cdot 0105 \, e \text{ mg.,}$$

z is the *electro-chemical equivalent* of the element.

A current of strength C (in ampères) will deposit a weight P per *second*:

$$P' = zC = \cdot 0105 \, eC \text{ mg.}$$

A current of strength C (in ampères) will deposit a weight P' *per hour*:

$$P' = 3600 \, zC = 37 \cdot 8 \, eC \text{ mg.}$$

One ampère hour (3600 coulombs) will liberate:

$37 \cdot 8$ milligrammes of hydrogen.
$37 \cdot 8 \, e$ milligrammes of any given element.

Those formulæ enable the deposit which would be produced by a given current in a given time to be calculated, or conversely the strength of current necessary to produce a given deposit in a given time.

Calculation of the electromotive force of polarisation of an electrolyte.—The electromotive force of polarisation of an electrolyte is a measure of the electro-chemical work done by the current in its decomposition.

The principle of the conservation of energy enables this e. m. f. to be calculated by equating the work done by the current in overcoming this polarisation e. m. f., and the mechanical equivalent of quantity of heat which the liberated element would disengage in recombining, so as to reform the original electrolyte.

Let E be the polarisation e. m. f. (in volts) of an electrolyte, Q the number of coulombs which has passed through it, the electro-chemical work of decomposition is:

$$\frac{QE}{9 \cdot 81} \text{ kgm.} \qquad (a)$$

If z be the electro-chemical equivalent of the liberated element (see page 212), the total weight liberated by Q coulombs will be equal to Qz.

---

* e the chemical equivalent $= \dfrac{\text{Combining weight or atomic weight}}{\text{Valency}}$.

Thus potassium a monad equivalent $= \dfrac{\text{Atomic weight}}{1}$.

Zinc, a diad, equivalent $= \dfrac{\text{Atomic weight}}{2}$.

Let H be the quantity of heat disengaged by 1 gramme of this element in combining so as to form the original electrolyte, then the heat disengaged by the weight Qz of this element will be QzH. As the mechanical equivalent of heat is ·424 kgm. per calorie (g.-d.) the heat disengaged by Qz grammes will be :

$$·424\, QzH. \qquad (\beta)$$

Equating (α) and (β) we get finally :

$$E = 4·16\, zH.$$

Electrolysis of water.—Applying the above formula to this case, as the heat disengaged by the oxydation of 1 gramme of hydrogen is 34450 calories (g.-d), and the electro-chemical equivalent of hydrogen is ·0000105, we get :

$$E = 4·16 \times ·0000105 \times 34450 = 1·5 \text{ volts.}$$

The polarisation e. m. f. in the electrolysis of water is thus 1·5 volts. This fact explains why one Daniell's cell is unable to decompose water, at least two in series being required.

Calculation of the e. m. f. of batteries.—The formula $E = 4·16\, zH$ enables the e. m. f. to be easily calculated for any voltaic combination when the nature of the chemical actions occurring in it and the heat disengaged by those actions are known.

The Daniell's cell.—Two distinct actions go on: 1st, solution of the zinc in sulphuric acid; 2nd, deposition of copper by the decomposition of copper sulphate.

1st. The heat disengaged by the solution of 1 gramme of zinc in sulphuric acid, H_1, is, according to *Julius Thomsen*, 1670 calories; the electro-chemical equivalent of zinc, $z_1 = ·0003412$; thus :

$$E_1 = 4·16 \times ·0003412 \times 1670 = 2·36 \text{ volts.}$$

2nd. The heat absorbed by the deposition of the copper H_2 is 881 calories per gramme, the electro-chemical equivalent of copper $z_2 = 1·21$ volts.

The e. m. f. of the Daniell cell is equal to the difference between the two, *i.e.* $2·36 - 1·21 = 1·15$ volts.

This theoretical value is not far from the practical value, 1·079 volts.

Electrolysis without polarisation.—When electrolysis is carried on with a soluble anode, if the bath be a solution of a *pure* salt of the metal, there is no polarisation, and the work done by the current is reduced to mere transport of matter from one plate to the other, which only requires a very small expenditure of electrical energy.* The expenditure of energy in the decomposition cell is thus practically reduced to the heating effect due to the passage of the current, and may be calculated by Ohm's law. If W be this work,

$$W = \frac{RC^2}{9\cdot 81} \text{ kgm. per second.}$$

R being the resistance of the bath in ohms, and C the strength of the current in ampères.

In practice the baths are never perfectly pure, and a certain amount of polarisation is produced in them which has to be taken into account.

ELECTRO-METALLURGY.

Under this title are included all operations in which a metallic deposit is formed by means of electrolysis.

The term "electrotype" is applied exclusively to those deposits which are so thick that they can be detached from the surface on which they have been deposited without losing their shape. Electrotypes are almost always of copper. Adherent deposits of this metal are used for coppering.

ELECTROTYPING.

Copper electrotyping.—For whatever purpose copper is to be deposited, the bath is always the same. It is thus prepared:

Bath.—A certain quantity of water is taken, to which from 8 to 10 per cent. of sulphuric acid is added slowly and by degrees, stirring well the whole time. As much sulphate of copper is then dissolved in this acidulated water as it will take up at the normal temperature, stirring well. The saturated bath ought to have a density of 1·21. It is always used cold, and must be kept saturated, either by the addition of fresh crystals or the use of soluble anodes. It must be used in vessels of porcelain, glass, hard faïence, or guttapercha. For large baths wooden vats are used, lined with a thin coating of guttapercha, marine glue, or varnished lead-foil. The vats should never be lined with iron, zinc, or tin.

---

* M. Lossier has calculated the energy absorbed by this transport by considering it as an induction effect produced by the movement in the fluid of polarised molecules.

Moulds.—Plaster of Paris is the substance which has been longest in use; but, as it is porous, it has to be made waterproof, which is a complication in its use. Moulds are now generally made of stearine, wax, marine glue, gelatine, guttapercha, and fusible alloy.

When the moulds are hollow, a skeleton of platinum wires is arranged in the interior. This is connected to the anode, and serves to direct the current, and so to render the deposit uniform in thickness. These wires are wound with a spiral of indiarubber to prevent their touching the walls of the mould. M. Gaston Planté uses lead instead of platinum for these wires, and has thus effected a great saving in cost.

When several things are being covered with metal at the same time, it is well to connect each one separately to the negative pole by an iron or leaden wire of appropriate thickness. This wire melts if there be any short-circuiting at its corresponding mould, and thus cuts it out of the circuit. The surface of the moulds is made conducting by pure plumbago, gilt or silvered plumbago, or bronze powder (a form of finely-divided copper, prepared by dropping granulated zinc into a solution of sulphate of copper). The mould is rubbed over with a clock-maker's brush or polishing brush. Wax requires very soft pencils. Moulds are also metalised by a wet process. A solution of nitrate of silver is brushed several times over the moulds, and reduced by the vapour of a concentrated solution of phosphorus in bisulphide of carbon. The wet method is best for very delicate objects, such as lace, flowers, leaves, moss, lichens, insects, etc. An agate cameo can be reproduced, without metalising, by simply winding a copper wire round it, and suspending it in the bath.

General management of baths and currents.—

When the solution is too weak, and the current is too strong, the deposit is *black;* when the solution is too strong, and the current is too weak, the deposit is *crystalline.* The metal is deposited in a sound, flexible state when the conditions are the mean between these two extremes: such a deposit was named by Smee *reguline.* The stratification of the liquid, and the circulation produced in the bath by the solution of the anode and the deposit on the cathode, produce long vertical lines, like notes of exclamation. The objects must be shaken about as much as possible, so as to keep the bath thoroughly homogeneous. Large baths are the best. A long distance between the anodes and cathodes produces a more regular deposit. It is especially necessary for small things; but it either decreases the rapidity of the deposition or requires a more powerful source of electricity. The same bath may be used for several objects at once, each one connected to a separate source of electricity, if one anode only be used, which is joined to the positive poles of all the sources. The surface of the anode generally ought to be the same as that of the cathodes: too small an anode weakens the solution; too large an anode strengthens it. Experience shows which effect it is desirable to produce in any particular case.

Copper clichés or electrotypes (*Stœsser*).—Wax moulds. Deposition requires twenty-four hours; mean thickness of deposit three-tenths of a millimètre, corresponding to a deposit of 25 grammes per square centimètre, or one gramme per hour per square centimètre. The strength of the current may be increased, so that the same thickness of deposit may be obtained in twelve hours, or less, without injuring the quality of the metal. Making the process last for twenty-four hours is, however, convenient, as the moulds may be prepared during the day and put in the bath at night.

Density of the current (*Sprague*).—The best bath for all objects not attacked by acid is:

Saturated solution of copper sulphate . . . 3 volumes.
Solution of sulphuric acid (one-tenth by volume). 1 ,,

The *density* of the current, that is, the current strength per unit of electrode area, may vary between certain limits which depend on the rate of work and the nature of the desired deposit. In this work we have taken as the unit of density, one ampère per square decimètre. One ampère deposits 1·19 grammes of copper per hour. The following are the results of Mr. Sprague's experiments:

Number of Experiment	Weight of Deposit per Hour per Square Decimètre, in Grammes.	Strength of Current per Square Decimètre, in Ampères.	Nature of Deposit.
1	·1	·085	Excellent.
2	·4	·342	Good tenacious copper.
3	3·	2·6	Magnificent.
4	12·	10·2	Very good.
5	50·	42·7	Powdery at the edges.
6	124·	106	Bad all round the edges.

ADHERENT DEPOSITS.

These are now deposited on all metals and of all metals. We will only describe the most important: coppering, brassing, gilding, silvering, and nickeling. All these deposits require a preliminary process of scaling

218   APPLICATIONS OF ELECTRICITY.

and cleaning to make the surface fit to receive the deposit, and to ensure as perfect adherence as possible between the two metals or alloys. Some hints on this process will be found in the Fifth Part.

Coppering is always performed by means of a bath of a double salt, either hot or cold; the composition of the bath varies with the nature of the metal to be coated. Below we give the formulæ recommended by an esteemed practical authority, M. Roseleur; but these formulæ differ immensely with different operators.

The copper acetate is dissolved in 5 litres of water, the ammonia and other substances in other 20 litres. They are mixed, and the fluid ought to be discoloured; if not, cyanide must be added until discoloration is produced, and then slightly in excess. The oldest baths work the best. Agitate the objects as much as possible. When the bath is too old, it may be refreshed by adding copper acetate and potassium cyanide in equal weights.

COPPERING BATHS (Roseleur).

The weights expressed in grammes are the quantities for 25 litres of water.

	Iron and Steel.		Tin, Cast Iron, and Zinc.	Small Zinc Object.
	Cold.	Hot.		
Sodium bisulphate	500	200	300	100
Potassium cyanide	500	700	500	700
Sodium carbonate	1000	500	—	—
Copper acetate	475	500	350	450
Ammonia	350	300	200	150

Brassing.—Hot (50° to 60° C.) for iron and zinc wire in bundles, and sham gold; cold for other objects.

For iron (cast and wrought), and steel.—Dissolve in 8 litres of soft water:

Sodium bisulphate	200 grammes.
Potassium cyanide (70 per cent.)	500 ,,
Sodium carbonate	100 ,,

Again, in 2 litres of water:

| Copper acetate | 125 grammes. |
| Neutral zinc chloride | 100 ,, |

Add the second fluid to the first. Avoid ammonia.

GILDING.

For zinc.—Dissolve in 20 litres of water:

Sodium bisulphate	700 grammes.
Potassium cyanide (70 per cent.)	1000 ,,

Again, in 5 litres of water:

Copper acetate	350 grammes.
Zinc chloride	350 ,,
Ammonia	400 ,,

Add the second fluid to the first, and filter.

Use a brass anode; add more zinc to produce a greenish deposit, more copper for a redder one. Too weak a current produces a red deposit, too strong a current a white or blueish-white deposit. Remedy by altering the battery-power, or using a copper or zinc anode. The density of the bath may be allowed to vary from 5° to 12° Baumé without injury.

Gilding (*Roseleur*).—Hot for small things, cold for large ones. *Bath of the double cyanide of gold and potassium.*—Cold.

Distilled water	10 litres.
Pure potassium cyanide	200 grammes.
Virgin gold	100 ,,

The gold transformed into chloride is dissolved in 2 litres of water, the cyanide in 8 litres; the two fluids are mixed, a change of colour takes place, and the mixture is boiled for half an hour. The richness of the bath is kept up by adding as required equal weights of pure potassium, cyanide and chloride of gold, a few grammes at a time. If the bath be too rich in gold, the deposit is blackish or dark red. If there is too much cyanide, the gilding goes on slowly, and the deposit is grey. The anode should be entirely immersed in the bath, suspended by platinum wire, and should be taken out when the bath is not working.

Hot gilding.—It is better to first copper objects made of zinc, tin, lead, antimony, and alloys of these metals. The following are the formulæ for other metals, the quantities being for 10 litres of distilled water:

	Silver, copper, and alloys rich in copper.	Cast and wrought iron, steel.
	grammes.	grammes.
Crystallised sodium phosphate	600	500
Sodium bisulphate	100	125
Pure potassium cyanide	10	5
Virgin gold transformed into chloride	10	10

Dissolve the sodium phosphate in 8 litres of hot water, allow it to cool; dissolve the chloride of gold in 1 litre of water, mix the second fluid slowly with the first; dissolve the cyanide and bisulphate in one litre of water, and add the third solution to the other two.

The temperature of the bath may vary from 50° to 80° C. A few minutes will produce thick enough gilding. A platinum anode is used; if it only just dips into the bath pale gilding is produced, if immersed very deeply the gilding is red. The bath may be refreshed by successive additions of chloride of gold and cyanide of potassium; but after long service it gives red or green gilding, according to whether it has been most used for gilding copper or silver. It is better to renew the bath than to refresh it.

Green, white, red, and pink gilding.—These different colours are obtained by mixed baths and currents of different strengths. *Green* is obtained by adding a very dilute solution of silver nitrate to the gold bath, *red* with a copper bath, and *pink* with a mixture of gold, silver, and copper baths.

Rate of deposition (Delval).—About 30 centigrammes per hour per square decimètre may be deposited from a bath containing 1 gr. of gold per litre, but this mean rate may be very much varied without harm.

Silvering.—A good bath for amateurs contains 10 gr. of silver per litre, and is thus prepared: dissolve 150 gr. of silver nitrate (which contains 100 gr. of pure silver) in 10 litres of water, add 250 gr. of pure potassium cyanide, stir till all is dissolved, and filter.

Silvering is generally done *cold*, except for very small objects. Iron, steel, zinc, lead, and tin (previously coppered) are best silvered *hot*. The articles after cleaning are passed through a solution of nitrate of binoxide of mercury, and are constantly agitated in the bath. When the current is too strong the articles become grey, turn black, and disengage gas. Use a platinum or silver anode in cold baths. Old baths are better than new. Baths may be aged artificially by adding 1 or 2 milligrammes of liquid ammonia. Silver baths are refreshed by adding equal parts of the silver salt and potassium cyanide. If the anode turns black the bath is poor in cyanide, if it turns white there is excess of cyanide; the deposition is then rapid, but does not adhere. When all is going on well and regularly the anode turns grey when the current is passing, and becomes white again when it is interrupted. The density of the bath may vary without injury between 5° and 15° Baumé.

Silver plating of forks and spoons (*Roseleur*).

1st. Boil them for a few moments in a solution of 1 kg. caustic potash in 10 litres of water, and wash in cold water.

SILVERING.

2nd. Pickle in water with one-tenth (by weight) of sulphuric acid.

3rd. Immerse for a few seconds in the following mixture:

Yellow nitric acid at 36°.	10 kg.
Common salt	200 gr.
Calcined tallow.	200 ,,

Wash quickly in plenty of water.

4th. Pass quickly through the following mixture:

Yellow nitric acid at 36°.	10 litres.
Sulphuric acid at 60°.	10 ,,
Common salt.	400 gr.

Wash quickly in perfectly clean water.

5th. Pass them through the following mixture **until they are quite white** (an immersion of a few seconds will do).

Water	10 litres.
Nitrate of binoxide of mercury	100 gr.
Just enough sulphuric acid to dissolve the binoxide.	

Wash in cold water.

6th. Place them in the bath, use a weak current; when sufficient silver has been deposited stop the current; allow them to remain for a few minutes longer in the bath, remove them, wash first in water, then in very dilute sulphuric acid; scratch-brush and burnish if necessary. The weight of silver deposited is 72, 84, or 100 grammes per dozen. The process lasts four hours, but a better quality of deposit is obtained by working more slowly.

Deposit per Hour per Square Mètre. In Grammes.	Nature of the Deposit.	Strength of Current in Ampères per Square Mètre.
140	Pin-holes.	35
200	Good deposit.	50
220	Deposit.	55

M. Delval gives, as a mean rate for a bath containing 30 gr. of silver per litre, 2 gr. per hour per square decimètre. This agrees with the above figures.

Nickeling (A. Gaiffe).—Nickel is principally deposited on copper, bronze, German silver, wrought and cast iron, and steel.

Cleaning from grease and pickling.—(*See* Fifth Part.)

Battery.—The handiest for amateurs is the bichromate bottle battery. The current is regulated by immersing the zinc more or less.

Bath.—Saturate hot distilled water with the double sulphate of nickel and ammonium free from oxides of alkaline, and alkaline earthy metals. The solution is:

Double sulphate of nickel and ammonium . .	1 part by weight.
Distilled water	10 ,,

Filter when cold.

Cell, and putting in the bath.—The best cells are of glass, porcelain, or earthenware; or of wood lined with waterproof varnish. Use a plate of nickel as a soluble anode, and suspend the articles by nickeled hooks. The objects are immersed for a moment in a bath of the same solution which has been used for pickling, washed quickly in common water, and then in distilled water, and quickly placed in the bath.

Rate of deposit.—M. Delval gives, as a mean rate for a bath containing 10 gr. of nickel per litre, 1·8 gr. per hour per square centimètre. This rate must not be much varied in order to get a good deposit in a bath of this strength.

Management of the current and time of the process.—If the current be too strong the nickel is deposited in a black or grey powder; one or two hours is long enough for a coating of mean thickness, five or six for a very thick coating.

Polishing.—Rub rapidly backwards and forwards on a strip of list fastened to a nail at one end and held by the other in the left hand; apply polishing powder and water to the list; hollow parts must be polished with pledgets of cloth fixed to handles. The polished objects are washed with water to remove the polishing powder and cloth fluff. In order to obtain a good polish the objects should be well polished before nickeling.

THERMO-ELECTRICITY.

It was discovered by *Seebeck* in 1821, that if the junction of two dissimilar metals be heated an e. m. f. is set up at the juncture. This e. m. f. is called *thermo-electromotive force*. The general group of electric phenomena produced by heating or cooling of junctions of dissimilar metals, and the passage of currents through such junctures, are known as *thermo-electric* phenomena. The thermo-electromotive force is constant when the temperature is constant and between certain limits, and for any pair of metals is proportional to the excess of temperature at the junction over the temperature of the rest of the circuit. The total thermo-electromotive force developed in any circuit is the algebraic sum of all the e. m. fs. developed at the different junctures. The thermo-electric power of two metals is the magnitude of the e. m. f. produced when there is 1° C. difference of temperature between the junctures.

The following table gives the thermo-electric powers in microvolts per degree C. of different metals, lead being taken as the standard. Bismuth being the most thermo-positive, and antimony the most thermo-negative metal, the current produced by a bismuth-antimony couple passes from bismuth to antimony across the juncture, and from antimony to bismuth through the external circuit.

TABLE OF THERMO-ELECTRIC POWER OF DIFFERENT METALS
With respect to Lead, at a Mean Temperature of 20° C. (*Matthiessen*).
(The electromotive forces are expressed in microvolts per degree centigrade.)

Commercial bismuth wire	+ 97	Pure antimony wire	− 2·8
Pure bismuth wire	+ 89	Pure silver	− 3
Crystallised bismuth in direction of axis	+ 65	Pure zinc	− 3·7
		Electro-deposited copper	− 3·8
Crystallised bismuth normal to the axis	+ 45	Commercial antimony wire	− 6·
		Arsenic	− 13·56
Cobalt	+ 22	Pianoforte wire	− 17·5
German silver	+ 11·75	Crystallised antimony, in direction of axis	− 22·6
Mercury	+ ·418		
Lead	0	Crystallised antimony, normal to axis	− 26·4
Tin	− ·1		
Commercial copper	− ·1	Red phosphorus	− 29·7
Platinum	− ·9	Tellurium	− 502
Gold	− 1·2	Selenium	− 807

Impurities have considerable influence on the thermo-electric power of metals; some alloys and some sulphides, such as galena (zinc sulphide), have a very high thermo-electric power.

Thermo-electric inversion.—Neutral point.

The thermo-electric power of metals is a function of the mean temperature of the junctions as well as of the difference of temperature between the junctions. The following figure enables the variations of this thermo-electric power to be studied. The abscissæ represent the mean temperatures in degrees C., and the ordinates the e. m. f. in microvolts. The distance between the two lines of two given metals at a given mean temperature shows the thermo-electric power of the two metals at that mean temperature. The lines are plotted, taking lead as a standard.

The point where the lines of metals cut each other is the *neutral point*. At the corresponding temperature the thermo-electric powers of the two metals are equal.

Beyond the neutral point the thermo-electromotive force changes its sign; thus the neutral point is also the *point of reversal*. *Tait* has shown that between 0° and 300° C. these lines are sensibly straight. The calculation of thermo-electromotive forces is thus reduced to the calculation of the areas of triangles or trapeziums. Let m be the distance in microvolts which separates the line of two metals at the mean temperature, $t_1 - t_2$ the difference of temperature in centigrade degrees, the e. m. f. would

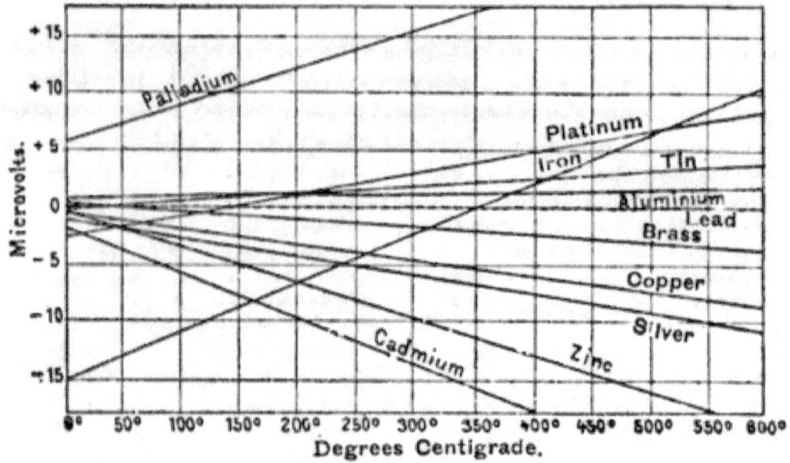

Fig. 44.—Diagram of Thermo-electric Powers.

then be $m(t_1 - t_2)$. Therefore, if the mean temperature is that of the neutral point or point of inversion, there will be no current produced, because m is nothing. This enables the neutral point of different metals to be determined. Thus we see that it is not sufficient to maintain a great difference of temperatures between the junctures to get a large e. m. f. It is also necessary to make a judicious choice of the metals and the mean temperatures, so as to keep as far as possible from the neutral point.

Formula for calculating thermo-electric power. —This formula, due to Tait, is a result of the graphic representation above. It is based on the supposition that the lines referring to the different metals are straight lines. This hypothesis has been verified experimentally between 0° and 400°.

Let k_1 and k_2 be the tangents of the inclinations of the lines of the two metals considered in relation to lead, t_1 and t_2 the neutral point of each in relation to lead, t_m the mean temperature of the two junctions, the mean ordinate m or thermo-electric power is given by the formula:

$$m = k_1 (t_1 - t_m) - k_2 (t_2 - t_m).$$

The total e. m. f. E_t is then:

$$E_t = m (t_1 - t_2).$$

TABLE FOR CALCULATING THERMO-ELECTRIC POWERS.

METALS.	NEUTRAL POINT WITH RESPECT TO LEAD IN CENTIGRADE DEGREES.	TANGENT OF THE ANGLE WITH LEAD k.
Cadmium	− 69	− ·0364
Zinc	− 32	− ·0289
Silver	− 115	− ·0146
Copper	− 68	− ·0124
Brass	+ 27	− ·0056
Lead	—	—
Aluminium	− 113	+ ·0026
Tin	+ 45	+ ·0037
German silver	− 314	+ ·0251
Palladium	− 181	+ ·0311
Iron	+ 357	+ ·042

Bismuth-copper couple (*Gaugain*) has been used as a standard in measurements of e. m. f. One of the junctions is at 0°, the other at 100°:

$$E_t = \frac{1}{182 \cdot 6} \text{ volt.}$$

Noe's pile.—One of the metals is German silver, the other an alloy of antimony and zinc. Each element in regular work gives an e. m. f. of $\frac{1}{16}$ volts and an internal resistance of $\frac{1}{40}$ ohm. The battery of 20 elements in series has therefore an e. m. f. of 1·25 volts and an internal resistance of ·5 ohm.

Clamond's pile.—Iron and an alloy of bismuth and antimony cast on the iron. A battery of 6,000 elements in series heated by coke produced a e. m. f. of 109 volts, and had an internal resistance of 15·5 ohms.

Thermo-electric batteries have as yet only been applied to the measurement of small differences of temperature and as standards of e. m. f. As yet the best of them transform less than one per cent. of the heat energy given out by the source of heat.

HEATING ACTION OF CURRENTS.

When a current passes through a conductor, it heats it. The quantity of heat produced by the passage of a current is given by the formula:

$$H = \frac{C^2 R t}{g J}.$$

H, quantity of heat produced in calories (g.-d.).
R, resistance of conductor in ohms.
C, strength of current in ampères.
t, time during which the current has passed in seconds.
J, mechanical equivalent of heat.

$$H = \frac{C^2 R}{4·16} \text{ calories (g.-d.).}$$

Minimum diameter for wires depends on the conductivity of the wire, its shape, the facilities it has for cooling, and the use to which it is applied. Conductors of platinum wire, used to light spirit lamps or gas, small incandescent lamps, etc., ought to become red-hot, but not to melt. The safety catches of cables for electric light leads ought to melt and automatically interrupt the circuit as soon as the strength of the current on the branch becomes double or triple that which it ought to be normally. The wires of dynamos and covered conductors ought never to reach the temperature at which the insulation would be damaged. In machines wound with wire not more than 2 millimètres in diameter, it is safe to allow a current to pass of 5 to 6 ampères per square millimètre, and only 3 ampères per square millimètre for wire of 5 millimètres in diameter. For lead-covered copper conductors, where the cooling is slow, not more than 2 ampères per square millimètre should be passed for a conductivity of 80 to 90 per cent., and currents less than 20 ampères.

Loss of energy in conductors.—The heat produced is calculated by the last formula. In all cases of transmission of energy to

HEATING EFFECTS OF CURRENTS.

a distance, it is advantageous to diminish this loss by reducing the resistance of the conductors as much as possible. In the following table the value of the energy wasted in the wire in the form of heat is given for currents from 1 to 100 ampères, and a resistance of 1 ohm, in calories (g.-d.), kilogrammètres per second, and horse-power.

Strength of Current in Ampères.	Calories (g.-d.) per Second.	Kilogrammètres per Second.	Horse-Power.
1	·24	·102	·013
2	·96	·408	·054
5	6·01	2·548	·034
10	24·03	10·2	·134
20	96·12	40·8	·536
30	216·2	91·7	1·223
40	384·48	163·1	2·144
50	601	255	3·4
60	865	367	4·892
70	1177	499	6·653
80	1538	652	8·576
90	1918	826	11·007
100	2403	1019	13·59

From this table, when the current is known, and the loss of current by heating of the conductor, which can be allowed, is known, the resistance, and hence the diameter, of the conductor can be calculated.

Heating of a conductor traversed by a current (*G. Forbes*).—Let C be the strength of the current which heats the conductor at a certain temperature, and D the diameter of this conductor. To heat another conductor placed under the same conditions to the same temperature C and D, it must be varied so as to satisfy the equation:

$$\frac{C}{D^{\frac{3}{2}}} = a;$$

a being a constant which depends on the temperature of the wire.

Heating of coils of the same dimensions wound with wire of different diameters (*G. Forbes*).— The length of wire wound on coils of the same dimensions is inversely proportional to the square of the diameter of the wire, and the resistance is inversely proportional to the 4th power of the diameter. In order that the two wires, under these circumstances, may be heated to the same temperature, the strength of the current C and the diameter of the wire D must satisfy the equation:

$$\frac{C}{D^2} = a;$$

a being a constant which depends on the temperature of the wire,

Heating of two similar coils by the passage of a current (*G. Forbes*).—Let C be the strength of the current passed through a given coil at a given temperature, C' the strength of the current which passes through a second similar coil, of which the linear dimensions are n times those of the first, and diameter of the wire n times that of the first wire, for the same temperature we have the following equation:

$$C' = n^{\frac{3}{2}} C.$$

This law, deduced by equating the heat produced by the current and the heat lost by the coil, has been confirmed by experiment.

ELECTRIC LIGHT.

Electrical energy used for electric lighting being now almost exclusively furnished by magneto and dynamo-electric machines, it is impossible to go into this important application of the heating power of currents until after having spoken of the mechanical generators of electricity. (*See* page 260.)

MECHANICAL GENERATORS OF ELECTRICITY.

We will not here discuss instruments based upon electrostatic actions. All mechanical generators of electricity based upon electro-magnetic action are composed of two parts, the field magnet and the armature. The field magnet produces the magnetic or galvanic field necessary for the production of the current which is produced in the armature, which is called an induced current. Every displacement of the armature relative to the field magnet produces an induced current. In all magneto and dynamo-machines this relative displacement is produced by a mechanical movement. The electrical energy developed is equivalent to the work expended in this movement. Induction coils, for example, are instruments in which the relative displacement of the field and the armature is produced without mechanical movement. They act as transformers, and are not mechanical generators of electricity, although the same principle underlies the action both of the coil and of the machines.

Work expended.—The work expended in driving a mechanical generator of electricity is measured either by a transmission dynamometer or by diagrams taken from the prime mover, taking into account the loss of work due to the friction in the prime mover itself. The work expended in driving a mechanical generator of electricity determines the choice of the prime mover which is to drive it. The power of electric generators varies from a few hundred kilogrammètres up to 400 horse-power.

Total electrical energy produced.—Let E be the e. m. f. developed by a generator, and C the strength of the current in its normal conditions of working. The total electrical energy W_t produced is:

$$W_t = \frac{CE}{9 \cdot 81} \text{ kgm. per second;}$$

or

$$W_t = \frac{CE}{746} \text{ horse-power} = \frac{CE}{736} \text{ chevaux-vapeur.}$$

The difference between the total electrical energy produced and the work expended represents losses due to friction, passive resistances, and complex secondary reactions, which are produced between the field magnets and the armature of the generators.

Available electrical energy.—Let E_a be the difference of potential at the terminals of the machines, *i.e.* between the ends of the external circuit, and C the strength of the current in the external circuit. The available energy W_a is:

$$W_a = \frac{CE_a}{98 \cdot 1} \text{ kgm. per second.}$$

$$W_a = \frac{CE_a}{746} \text{ horse-power} = \frac{CE_a}{9 \cdot 81} \text{ chevaux-vapeur.}$$

Heating of the machine.—The difference between W_t and W_a, expressed in kilogrammètres or horse-power, represents the work transformed into heat in the machine, or its heating. This being lost work, it is advantageous to diminish it as much as possible, which may be done by reducing the internal resistance of the machine in relation to the resistance of the external circuit.

When the internal resistance R_i of the machine is known, its heating is calculated by the formula:

$$Heating = \frac{C^2 R_i}{9 \cdot 81} \text{ kgm. per second} = \frac{C^2 R_i}{746} \text{ horse-power;}$$

or, $Heating = \dfrac{C^2 R}{424}$ calories (kg.-d.) per second.

Relation between the external and internal resistance.—Calling R_e the external resistance, and R_i the internal resistance, then

$$R_e = \frac{R_i}{3} \text{ or } \frac{R_i}{2},$$

when the machines are under bad working conditions;

$$R_e = 10 \text{ to } 40\, R_i,$$

when the machines are under good working conditions.

Classification. — The principal characteristics of electrical generators which may be used as bases for classification are (*a*) the nature of the currents produced, (*b*) the nature of the field magnet, (*c*) the form of armature. As well as these principal characteristics, other secondary ones exist, such as the nature of the moving part, the power of the machine, the presence or absence of iron in the armature, etc. We will only consider here the first class of characteristics.

a. THE NATURE OF THE CURRENTS PRODUCED.—Induced currents are always alternating by their very nature;[*] some machines re-reverse them, others make them sensibly continuous; hence three classes of machines may be formed, according to the character of the currents which they give.

(1) *Alternating machines.*—The currents produced are collected in the same condition as the coils produce them, and are thus reversed as often as 30,000 times per minute.

(2) *Re-reversed machines.*—A commutator re-reverses the currents developed in the armature each time that they are about to change sign. The current then becomes zero at each commutation. The type is the Siemens H armature machine.

(3) *Continuous current machines.*—The armatures which are split up into

---

[*] Except in those machines which are very wrongly called *uni*-polar, which have not yet come into practical use; they may be considered as practically continuous.

sections are joined to a commutator, which produces a great number of partial commutations. The strength of the current is, as it were, sinuous, and becomes more and more closely a straight line or steady current as the splitting up of the armature is greater. They may be considered as practically continuous. If a telephone, however, be placed between the terminals of the machine, it will show by its vibrations that the current is not altogether steady.

b. NATURE OF THE FIELD MAGNETS. — This character allows the machines to be divided into two classes:

(1) *Magneto-electric machines.* — The field magnet is a permanent magnet.

(2) *Dynamo-electric machines.* — The field magnet is an electro-magnet.

c. FORM OF THE ARMATURE. — The following forms may be distinguished:

(1) *Ring :* Elias Pacinotti, Gramme, Schuckert, Brush, etc.
(2) *Drum ;* Siemens, Edison, etc.
(3) *Pole :* Lontin, Niaudet, Wallace-Farmer, etc.
(4) *Disc :* The " Arago " machine, Ferranti-Thomson.

The armature, according to its shape, is called coil, ring, drum, etc.

Modes of excitation of dynamo-electric machines.

—To excite a machine is to supply it with the electrical energy necessary to maintain the magnetism of the field magnets. These are the methods employed up to the present time:

Separate excitation. — The field magnets are supplied by a distinct current furnished by a separate machine, which is called the exciting machine. The exciting machine may excite many dynamos at the same time, which produces a certain homogeneity in their magnetic fields, and consequently in their production and in their power.

Excitation in circuit. — The current produced by the machine itself passes through the field magnets, and keeps up their magnetism. Such machines are called *series dynamos.*

Shunt excitation. — The field magnets are arranged as a shunt between the brushes. The current produced by the armature is divided between the external circuit and the field magnet circuit (indicated by Wheatstone). Such machines are called *shunt dynamos.*

Double circuit excitation (indicated for the first time by Brush).—The field magnets are wound with two wires, one receiving the current from a separate machine, or arranged as a shunt between the brushes; the other wire being in the main circuit. Applied by M. Marcel Deprez to his system of distribution. This method enables the production of electrical energy to be made more or less proportional to the consumption in the external circuit. A great number of such arrangements may be invented.

Nature of the machines.—In normal conditions the working of the machine develops an e. m. f. of E, and gives out a current of strength C.

As a remnant of the old misleading nomenclature, machines are still sometimes divided into two classes. Tension machines, those in which E is very large; and quantity machines, those in which C is very large. But now machines are constructed which run through the whole scale of e. m. fs. from a fraction of a volt up to more than 3,000 volts, and the whole series of current strengths from a few milliampères up to 1,000 ampères and more.

Field magnets.—The field magnets ought to be as powerful and as massive as possible, in order to keep the magnetic field constant in spite of the rotation of the armature, and so as not to reach the point of magnetic saturation too soon. The field magnets ought to be made of as soft iron as possible, because its point of saturation is higher, and the poles ought to be as close to the armature as the design of the machine allows. Points and edges should be avoided. Care must be taken that the cast-iron bed plate of the machine does not disturb the distribution of the magnetic field by setting up a sort of magnetic short circuit between the poles.

Armatures.—Internal iron in the armature is of advantage so far as it concentrates the magnetic field, but difficulties of construction are found if it is to be kept fixed so as not to lose the work due to its changes of magnetisation when it turns with the armature, as for example in the gramme ring. A revolving armature ought to be of as soft iron as possible, and split up into plates, wires, etc., separated by varnish, asbestos, paper, or thin leaves of mica so as to hinder the development of Foucault currents. Also, to avoid heating, abrupt changes of polarity of the armature should be avoided. The winding should be arranged so as to reduce to a minimum those parts of the wire which do not come under the direct and useful action of the magnetic field. Wire of the highest conductivity possible should be chosen. (At least 96 per cent. of conductivity.) When an armature has poles the wire ought to be wound as near these poles as possible, because it is at this point where the variations are most sensible. The armature ought to be easily ventilated, and the coils perfectly insulated with substances which are not easily melted, such as asbestos, mica, etc.

Conditions to be aimed at in the construction of a powerful machine.—The current developed in a conductor which is passing through a magnetic field is greater the more lines of force the conductor cuts through per unit of time. Consequently, to build a powerful machine it is necessary, first, to give the conductor a high velocity of translation; secondly, to move it in an intense magnetic field; thirdly, to have as many turns as possible on the armature; fourthly, to diminish its electrical resistance as much as possible. The first condition is limited by mechanical considerations. In practice, 20 to 25 mètres per second is rarely exceeded for the middle part of the armature. The second condition explains the use of dynamo-electric machines, which give an intense magnetic field, and which for the same speed require fewer turns of wire on the coil than magneto-electric machines in order to develop the same e. m. f. It is thus possible to increase the thickness of the wire and diminish its length, and thus decrease the internal resistance of the machine. The advantage obtained more than compensates for the additional expenditure of energy in exciting the field magnets. The third condition shows that the number of turns of wire must be greater as the e. m. f. has to be greater, which forces us to diminish the diameter of the wire, so that it may occupy the same space. It is thus not *because* the machines have a high resistance, as is sometimes said, that they have a high e. m. f., but because it is impossible to put a large number of turns of wire into a given space without using fine wire, and thus introducing a high resistance. If high resistance were a necessary and sufficient condition for high e. m. f., coils might be made of German silver or platinum wire; but, in practice, copper wire of the highest conductivity is always selected. The fourth condition to be fulfilled further justifies this choice, because Ohm's law shows that the strength of a current diminishes when the total resistance of a circuit increases.

Influence of the thickness of the wire.—Other things being equal, the e. m. f. of the machine does not increase in proportion to the diminution of the section of the wire, as some authors have considered. If a given machine, for example, develops 200 volts with a wire two millimètres in diameter, it would not necessarily develop 800 volts with a wire one millimètre in diameter, because the thickness of the insulating coating is greater in relation to the diameter of the wire as the wire is thinner. The second machine would thus not contain four times as many turns as the first, either on the armature or on the field magnets. The e. m. f. would diminish then for two reasons: because the magnetic field would be less intense, and the number of turns of wire on the armature less than the theoretical proportionality requires. Lack of wire in the field magnets diminishes the e. m. f. of a given machine as the strength of the current given out increases. This fact is a result of the reciprocal reactions of the magnetic and galvanic fields.

A magnetic field costs the less to produce, as the electro-magnets are of larger dimensions (*Edison*). This is a consequence of the general laws of electro-magnets.

Influence of speed on the work absorbed.—With a constant magnetic field the work absorbed is sensibly proportional to the square of the speed of rotation, and the e. m. f. proportional to the speed, when the external circuit remains constant. In the case of series or shunt dynamo machines, the phenomena is much more complicated, because of the progressive increase of strength of the field with the increase of the current.

Characteristic (*Marcel Deprez*).—The curve given by the relation between the e. m. f. of a given dynamo revolving at a given speed and the strength of the current which it produces when the resistance of the circuit is varied. To construct the curve, the field magnets are separately excited for each current strength, C corresponding to an e. m. f. E. The current strengths are made abscissæ, and the e. m. fs. ordinates.

Cabanellas has justly observed that the curve traced by this method is wrong in all cases in which the magnetic field is modified by the electric field. The curve thus obtained varies with the dimensions of the machine and the relation of its different parts. Generally it is more or less parabolic in form, and is very rarely approximately a straight line.

Critical speed.—The speed at which on a given circuit a dynamo begins to give out a sensible current: a very slight decrease of speed practically reduces the current strength to zero, whilst a very slight increase in speed produces a very large increase of current strength. This critical speed has been made the basis of some systems of regulating dynamos and of governing motors.

Lead of the brushes (*A. Breguet*).—In magneto-electric machines, or separately-excited dynamos acting as generators of electricity, the angular lead of the brushes should increase with the speed of rotation. For series machines the brushes require but a small lead so long as the field magnets are not saturated; and after saturation, especially with weak field magnets, the lead may be as much as $70°$.

In dynamo-electric machines acting as motors, the lead should be against the direction of the movement of rotation. The angle of the lead ought to be greater as the magnetic field is weaker and the current stronger. These facts are consequences of the reciprocal action of the magnetic and galvanic fields of the field magnets and the armature. The lead necessary for any given set of circumstances and the change of lead necessary on their variation vary very much with different types of dynamos, some requiring much more change than others. It is, however, well always to have some simple means of changing the lead whilst the machine is running, and to shift the lead whenever much sparking is observed at the brushes until the sparking is reduced to a minimum.

Action of the iron ring in Gramme machines (*A. Breguet*).—(1) The retardation of magnetisation and demagnetisation makes it necessary to give the brushes a lead in the direction of the rotation of not more than 10° for the highest speeds.

(2) The presence of the ring increases the intensity of the field and reduces its distortion, and consequently the lead of the brushes.

The soft iron armature of the Gramme machine reinforces the magnetic field in the part in which the wires of the movable circuit move, and protects the internal parts of the coils from the normal action of the lines of force of the field by acting as a magnetic screen.

Size of wire for dynamo-electric machines.—In the Gramme machines built by Sautter and Lemonnier the following proportions are used:

DETAILS.	TYPES.				
	M.	AG.	CT.	CQ.	DQ.
Armature.					
Diameter of wire in millimètres	1·2	1·8	2·8	3·65	4·3
Strength of current given out	13·5	24·5	48	65·	70
Strength of current passing through the wire	6·75	12·25	24	32·5	35
Section in square millimètres	1·13	2·54	6·16	10·46	14·52
Current strength per square millimètre	6	4·8	3·9	3·1	2·4
Field Magnets.					
Diameter of wire	1·8	3·4	3·4	3·4	3·8
Section in square millimètres	2·54	9·03	9·08	9·08	11·34
Strength of current	6·75	12·25	48	65	17·5
Strength of current per square millimètre	2·6	1·3	5·3	7·2	1·6

In the case of the armature the law is very clear, so many fewer ampères per square millimètre of section must be passed as the wire becomes larger, because with large wire the cooling is slower. For field magnets the figures vary much with the method of connection.

Maintenance of the brushes and commutators.—The brushes should press moderately on the commutators, and their lead be so arranged as to give the least sparking; that is to say, contact of the brush with the armature should be opposite the neutral point. The brushes should be pushed forward slightly as they wear.

If some of the wires get fused together by sparks, they must be carefully separated. If the wires become detached and turned up, they must be bent into their proper shape with a pair of flat-jawed pincers. They should be cleaned with alcohol from time to time. The commutators may be lubricated with a piece of rag very slightly moistened with oil; but care must be taken that not too much oil be applied. Sometimes they are rubbed with mercury, but this is a very bad practice. The brushes should be smoothed from time to time by rubbing them with emery-paper. Never break the circuit by throwing off the brushes when the machine is running. The machine ought always to pull on the brushes; that is the simple practical way of determining which way the machine ought to turn.

Best working conditions for machines.—There is not space to give here a description of all the different magneto and dynamo-electric machines which are now used in practice; we can only point out the best conditions of working of those most in use, or, at least, of those of which we have been able to collect the elements, but throwing the whole responsibility of the figures themselves on to the inventors or experimenters. The reader will rather find here useful information on the mean conditions of working, than on the relative value of different machines. Machines having been, up to the present time, for the most part specially constructed for electric lighting, a good deal of information on dynamos will be found in that part of the present work which deals with that subject.

CONTINUOUS CURRENT MACHINES.

A Gramme machine (*Breguet's* construction). — Series dynamo:

Resistance of the ring cold . . . ·47 ohm.
,, ,, field magnets coil . . ·37 ,,
Total resistance cold . . . . 1·14 ,,
,, ,, hot . . . . 1·2 ,,

CONTINUOUS CURRENT MACHINES. 237

Normal speed	900 revs. per minute.
Strength of current	25 to 30 ampères.
Electromotive force	80 volts.
Difference of potential at terminals	55 volts.
Field magnets saturated at	18 ampères.

Heinrich's machine.—U-shaped Gramme-ring dynamo. Experiments made on the type for 3 lamps in circuit (arc; carbons 13 mm. in diameter) gave the following results (*Kempe, Preece,* and *Stroh*):

Total resistance of machine	1·83 ohms.
Resistance of ring	·85 ,,
Electromotive force	130 to 150 volts.
Strength of current	33 to 38 ampères.
Revolutions per minute	850
Mean diameter of ring	20 centimètres.

Against carbon resistances the same machine gave the following results:

Number of Revolutions.	External Resistance.	Strength of Current.	Electromotive Force.	Work in Horse-Power (Calculated).
700	2·1	36·4	143·3	7
800	2·6	33·7	149·3	6·5
900	4·3	26·3	160	5·5
900	7·3	15·7	143·3	3
1000	7·3	17·7	161·6	4

Gülcher machine.—Six-lamp machine, feeding six arc lamps in parallel arc at speed of 640 revolutions; field magnets in the circuit:

	Cold (16° C.)	Hot after running some hours. (31° C.)
Resistance of field magnets	·126 ohm.	·129 ohm.
,, ring	·133 ,,	·136 ,,
Total resistance	·259 ,,	·265 ,,
Difference of potential at terminals		60 volts.
,, ,, at brushes		70·22 ,,
Total strength of current		80 ampères.

	Hot after running some hours. (31° C.)
Work absorbed by generator	2·86 horse-power.
,, ,, external circuit	7·13 ,,
,, ,, friction	·5 ,,
Total work absorbed	10·49 ,,

The machine converts 68·2 per cent. of the work put into it into electrical energy available in the external circuit, and produces 75 carcels per horse-power at an angle of about 35° (*Gulcher*).

SCHUCKERT MACHINE (F. *Uppenborn*).

Name of Type.	Strength of Current in Ampères.	Volts at Terminals.	Number of Turns per Minute.	Work Absorbed in Horse-Power.
Monophote.				
$EL_{\frac{1}{2}}$	7	50	1,300	1
EL_1	20	50	1,100	2
EL_2	36	50	1,000	
EL_3	50	50	950	5·5
Polyphote and Transmission of Power.				
TL_1 2 lamps	8	100	1,200	2
TL_2 3 ,,	8	150	1,150	3
TL_{2a} 4 ,,	8	200	1,100	4
TL_5 6 ,,	8	300	1,000	5·5
TL_4 9 ,,	8	450	950	8
TL_5 14 ,,	8	750	930	12
Incandescent (*compound-dynamos*)				
$JL_{\frac{1}{2}}$ 5 lamps	3·6	110	1,500	1
JL_1 12 ,,	8·6	110	1,300	2
JL_2 18 ,,	13	110	1,200	3
JL_{2a} 25 ,,	18	110	1,100	4·3
JL_3 35 ,,	25·2	110	1,000	6
JL_4 50 ,,	36	110	900	8·5
JL_5 72 ,,	52	110	750	12
JL_6 40 ,,	288	110	500	60

SIEMENS MACHINES.

Schuckert electroplating machines.—The nickel-plating machines (shunt dynamos) give 4 volts at the terminals, and from 90 to 1,400 ampères, absorbing from 1 to 12 horse-power. The copper-plating machines give 2 volts at the terminals, and from 200 to 2,800 ampères absorbing the same power. The 200-ampère machine deposits 225 gr. of nickel per hour on a surface of 2·5 sq. mètres. The 200 ampère copper-plating machine deposits 816 gr. of copper per hour on a surface of 83 sq. decimètres, by arranging four baths in series.

SIEMENS MACHINES (1883).

Type of Machine.	Number of incandescent lamps of 45 volts and 1·34 ampères.	Number of Watts.			Total number of watts developed.	Percentage available in the external circuit.	Total weight of copper in the machine in kilogrammes.	Number of watts disposable per kilogramme of copper in the external circuit on the machine.	Mean speed of circumference in mètres per sec.
		In the coil.	In the field magnets.	In the external circuit.					
(1)	(2)	(3)	(4)	(5)	(6)	(7)	(8)	(9)	(10)
S D 5	12	12	248	796	1,164	68	19·5	41	9·5
S D 7	25×2	308	370	3,316	3,994	83	51·25	64	13·75
S D 7	40	328	233	2,653	3,214	82	64	41	10·75
S D 2	30×2	536	319	3,980	4,835	82	119	33·4	9·25
S D 2	60	526	326	3,980	4,832	82	118·5	33·5	9·5
S D 1	60×2	803	532	7,959	9,294	85	264	30·1	10·25
S D 1	50×3	615	1,200	9,949	11,764	84·5	274	37·7	13
D S D 00	150×2	2,080	2,562	19,890	24,532	91·7	389	51	10·75
B 1 .	400	1,654	2,665	26,532	30,851	86	368	72	13·75
W 3 .	60	193	456	3,979	4,868	81·7	103	38·5	18·25
D 5 .	—	97	143	—	—	—	—	—	7·75
W 6 .	90	335	595	5,969	7,240	82	141	42·5	20·5
D 6 .	—	121	220	—	—	—	—	—	8·15
W 2 .	120	630	633	7,959	9,523	83·5	163	49	22
D 6 .	—	107	194	—	—	—	—	—	9·25
W 1 .	200	1,357	1,257	13,266	16,034	82·7	237	56	24
D 7 .	—	110	94	—	—	—	—	—	10·75
W 5 .	400	1,034	1,496	24,030	26,777	90	227	106	28·9
D 7 .	—	117	160	—	—	—	—	—	11·15
W 0 .	500	1,357	2,375	33,165	37,460	88	384	86·5	32·5
D 2 .	—	177	206	—	—	—	—	—	9·5

Continuous current Siemens machines.—Normal working conditions:

Type.	Number of Revolutions per Minute.	Volts.	Ampères.
D⁰	750	95	108
D¹	450	84	70
D⁵	1200	50	22
D⁶	1100	67	20
D⁷A	1250	129	20
D⁷B	1000	62	37
D⁸A	800	336	8·8
D⁸B	1000	224	20
D⁸C	650	78	37

Edison machine.—Continuous current shunt machine, giving 110 volts at the terminals and current strength proportional to the number of A lamps arranged in parallel arc. They are built at Paris (1882) of the following types:

Type.	Number of A Lamps.	Weight of the Armature.	Weight of the Field Magnets.	Total Weight.	Width of Pulley.	Diameter of Pulley.	Number of Revolutions.	Resistance of the Armature.	Resistance of Field Magnets.	Work in Horse-Power at the Pulley.
E	15	108·84	226·75	335·5	76·2	127	2200	·36	90	3
Z	60	136·65	859·38	1233	152·4	254	1200	·138	38	10
L	150	226·75	1673·41	2580	228·6	355	900	·071	19	17·7
K	250	317·45	2428·49	3814	—	—	900	·032	13	34·4
C	1200	6031·55	13151	28707	—	—	350	·0038	25	123

Edison machine (1881).—Type for 1,000 lamps of 16 candle-power, or A lamps.

Armature.—108 bars of copper, 2,000 iron discs forming the core: diameter, 71 centimètres; length, 1·5 mètres; surface velocity, 11·3 mètres per second; resistance of the armature, ·0005 ohm.

Field Magnets.— 12 bars of iron 2·4 mètres long, and 12 coils in two parallel circuits, of which the resistance is 21 ohms, arranged as a shunt between the brushes. The driving engine is 128 horse-power nominal.

Edison-Hopkinson machine, 1883.—The field magnets are shortened and the relative dimensions of the parts changed. The two field magnets joined up in series have a resistance of 35·5 ohms cold, and 37 ohms hot. The resistance of the armature is ·026 ohm cold, and ·0325 ohm hot; its diameter is 22·5 cm. without the wire, and 26·3 cm. with the wire; the mean length of the armature is 108 centimètres. The following table shows the principal results obtained by *Mr. Sprague* in three experiments on this machine. The coefficient of transformation is the relation between the work absorbed and the work transformed into electrical energy. The commercial efficiency is the relation between the work absorbed and the electrical energy available in the external circuit. The machine can supply about 10 B lamps per horse-power absorbed.

Particulars.	1	2	3
Number of revolutions per minute	1,081	1,157	1,179
Number of B Edison lamps supplied	199	199	237
Strength of current in ampères:			
In the field magnets	2·68	2·92	2·95
In the armature	112·6	123·1	144·6
Total	115·24	126	147·6
Electromotive force in volts	103	112·1	114·1
Difference of potential at terminals	99·3	108	109·3
Electrical energy in horse-power:			
In the field magnets	·36	·42	·43
In the armature	·58	·69	·95
In the lamps	14·97	17·81	21·18
Total	15·91	18·92	22·56
Work absorbed	16·63	20·12	23·79
Total work furnished	17·34	20·88	24·56
Coefficient of transformation	95·7	94	94·8
Commercial efficiency	85	85	86

Bürgin machine.—48 coils of wire, each wire 48 feet long, covered with cotton, ·065 inch in diameter. Internal resistance from brush to brush, 1·6 ohms. Field magnet: 1·2 ohms, wire ·141 inch in diameter. At 1,500 revolutions the e. m. f. is 195 volts. The velocity of the mean circle is 2,550 feet per minute.

242   APPLICATIONS OF ELECTRICITY.

DETAILS.	16 SHUNTS OF 3 LAMPS IN SERIES.	15 SHUNTS OF 2 LAMPS IN SERIES.	20 SHUNTS OF ONE SINGLE LAMP.
Speed	1540 to 1560	1260 to 1275	1290 to 1340
Electromotive force (volts)	186	135	138
Strength of current (amp.)	15	13·26	20·21
Engine power (from diagram)	5·05	4·05	5·22
Work expended in h.p.	3·67	2·67	3·84
Electrical energy produced in h.p.	3·3	2·4	3·7
Internal work (per cent.)	23	28	42
Loss in conductors	16	19	29
Energy utilised in the lamps	60	52	29
Total work in external circuit	76	69	58
Number of lamps per h.p.	13·1	11·3	5·2

Brush 16-light machine (arc lights).—(1879, *Brush*.)

Mean speed in revolutions per minute	770
Electromotive force	839 volts.
Strength of current given out	10·04 ampères.
Resistance of machine from terminal to terminal	10·55 ohms.
,, external circuit	72·96 ,,
Total resistance	83·51 ,,
Total work expended	15·48 h.p.
Work utilised by the machine	13·78 ,,
Electrical energy produced	11·28 ,,

87 per cent. of the total electrical energy produced by the machine appears in the external circuit; the machine transforms 81·89 per cent. of the energy which it absorbs into electrical energy. The lamps work with a difference of potential at their terminals of 45 volts. The shunt-regulating magnet absorbs about $\frac{1}{100}$ of the current. (The diameter of the carbons and the candle-power of the lights has not been given by the experimenters.)

Elphinstone-Vincent machine (1882).—Armature without iron. Resistance of armature, ·0374 ohm. The field magnets form two separate circuits of 8·5 ohms each. They are always arranged so as to make the machine a shunt dynamo, and are grouped either parallel (R = 4·25 ohms), or in series (R = 17 ohms) by turning a commutator. At 868 revolutions, field magnets parallel (R = 4·25) on an external resistance of ·4 ohm, the machine gave a current of 180 ampères and 72 volts at the terminals. At 855 revolutions it fed 144 shunts of two Swan 20-candle lamps in series each ($C = 1{\cdot}32$ ampères). With the field magnets in series ($R = 17$ ohms), and at a speed of 1,050 revolutions, it fed 152 Swan lamps in 76 shunts of two lamps in series each.

ALTERNATING CURRENT MACHINES.

Ferranti-Thomson machine (1882) (*Phillips*).—Alternating currents, revolving armature without iron, rotating at 1,900 revolutions.

Resistance of field magnets	2·5	ohms.
,, of exciting machine	·5	,,
Total resistance of exciting machine circuit	3	,,
Exciting current	22·5	ampères.
Strength of useful current	156	,,
Total electromotive force	125	volts.
Resistance of armature	·0265	ohm.
External resistance	·7735	,,
Total	·8	,,
Total electrical energy given out by the machine	1947	kgm. per second.
Electrical energy expended in excitin the field magnets	152	,, ,,
Total electrical energy	2099	,, ,,
Electrical energy available in the external circuit	1874	,, ,,
Electrical efficiency	·89	(89 per cent.).

The total weight of the machine is 550 kilogrammes; it feeds 300 Swan lamps, each requiring 41 volts between terminals, and a current of 1·3 ampères, arranged in 100 shunts of three lamps in series.

Alternating current Siemens machine.— At the normal speeds of working, and with suitable exciting circuits, each coil can develop an e. m. f. of 50 volts and give a current of 12 ampères. The power of each machine depends on the number of coils; and the character of the current given out, on the way in which they are grouped, whether in one or several circuits, parallel, or in series, etc.

The W^2 Siemens machine feeding 12 arc lamps arranged in three circuits of four lamps in series, gave the following results to the Commission at the Paris Electrical Exhibition in 1881:

	Generating machine.	Exciting machine.
Number of revolutions per minute	620	1230
Work expended in chevaux de vapeur	13·79	2·6
Strength of current in ampères	12·8	16
Diameter of wire on field magnets in mm.	3·5	3·5
,, ,, armature ,,	2·5	2
Fall of potential in the arcs	55 volts.	
Electrical work in the arcs	11·31 chevaux-vapeur	= 11 h.p.
Total electrical work	15·26 ,, ,,	= 15 h.p.

The machine transforms 93 per cent. of the mechanical work put into it into electrical energy, and there appears as available electrical work in the external circuit, 69 per cent. of the mechanical work expended, and 74 per cent. of the total electrical energy produced. With carbons 10 mm. in diameter, each lamp gives 39 carcels (mean spherical intensity) or 33 carcels per mechanical h.p., or 41·4 carcels per electrical h.p., and 33 carcels per ampère.

Méritens' machine (1880).—A magneto machine with five discs and sixteen coils on each disc. Each disc feeds one Berjot lamp with 20 mm. carbons. The sixteen coils are grouped in four groups parallel, each group consisting of four coils in series.

RESULTS OF THE COMMISSION AT THE PARIS EXHIBITION OF 1881.

Number of revolutions per minute	874
Total work put in	12·28 chevaux-vapeur, = 12 h.p.
,, ,, on open circuit	4·55 chevaux-vapeur, = 4·47 h.p.
Total work in the arcs	8·4 chevaux-vapeur, = 8·26 h.p.
Resistance of each disc	·18 ohm.
Strength of current	35·8 ampères.
Difference of potential at terminals of each lamp	36 volts.

The machine transforms 85 per cent. of the mechanical work put into it into electrical energy, and gives 68 per cent. of the work expended in available electrical energy in the external circuit. It works five lamps with a mean spherical luminous intensity of 150 carcels each, or 60 carcels per horse-power of electrical energy or arc horse-power, and 3·5 carcels per ampère.

ALTERNATING CURRENT MACHINES.

Measurement of current strength and electromotive force of alternating current machines.— The most elegant and most accurate mode of determining the working conditions of alternating current machines is by using Mascart's quadrant electrometer, by an idiostatic method due to *Joubert* (1880).

Let A be the potential of one pair of quadrants, B the potential of the other pair; let the needle be joined to the A pair of quadrants: then, if d be the deflection and k a constant depending on the instrument and the adjustment of the bifilar suspension, the general formula gives,

$$d = \frac{k}{2}(A-B)^2.$$

Determination of the constant k.—Each pair of quadrants is connected to one of the poles of a well-insulated battery of n elements in series, each element having a known e. m. f. e (say Daniell cells); the deflection is noted, and from it is deduced,

$$k = \frac{2d}{n^2 e^2}.$$

If e be expressed in volts, the instrument gives $(A-B)$ in volts and C in ampères.

Determination of current strength.—A known resistance R (in ohms), so arranged as not to be liable to self-induction (straight wire, carbon rods, or doubled and coiled wire) is introduced into the circuit, and the two pairs of quadrants are joined to the two ends of this resistance, acquire their potentials and deflect the needle through d; then,

$$d = \frac{k}{2}(A-B)^2.$$

A and B being the potentials at the ends of the resistance R, applying Ohm's law we get for the current strength c,

$$c = \frac{1}{R}\sqrt{\frac{2d}{k}}.$$

Let

$$\sqrt{\frac{2}{k}} = a.$$

The formula now becomes,

$$c = \frac{a\sqrt{d}}{k}.$$

The deflection is independent of the sign of the difference of potential.

With alternating currents the period of which is very small compared with the period of oscillation of the needle, the needle being always solicited in one direction takes up a fixed deflection proportional to the mean of the successive values of the square $(A-B)$. This method therefore gives the *mean* current strength, which is what would be given by the electro-dynamometer or a calorimetric method.

Determination of the difference of potential at the terminals of the machine, or at the terminals of a lamp or other apparatus using the current.—The instrument again gives the mean value of the difference of potential $(A-B)$ between different points of the circuit. Let d' be the deflection obtained when the quadrants are joined to two given points, the formula becomes,

$$(A-B) = \sqrt{\frac{2d'}{k}}.$$

The electrometer is the only known instrument which enables the mean difference of potential between two points of a circuit traversed by an alternating current to be found directly.

Determination of the energy absorbed.—The current strength C is measured in ampères, and the difference of potential E between the two given points of the circuit is measured in volts. Then for the energy W consumed, we have

$$W = \frac{CE}{9 \cdot 81} \text{ kgm. per second} = CE \times 1 \cdot 35 \text{ foot-lbs. per second.}$$

If d be the deflection of the electrometer connected to the two ends of the known resistance R, and d' be the deflection between the two ends of the consuming part of the circuit, we get,

$$W = \frac{a^2}{9 \cdot 81 R} \sqrt{dd'} \text{ kgm. per sec.} = \frac{1 \cdot 35 \, a^2 \sqrt{dd'}}{R} \text{ foot-lbs. per second.}$$

Joubert remarks that in the case of alternating currents the foregoing equations give the value of the expression,

$$\frac{1}{Rt} \sqrt{\int (A-B)^2 dt \times \int (A'-B')^2 dt}.$$

t being any time large in comparison with the period of alternation, but that the true expression for the work is,

$$W = \frac{1}{Rt} \int (A-B)(A'-B') \, dt.$$

In practice these two integrals do not differ sensibly from each other. The following is a method due to *Potier* (1881), which gives the exact value of the work, though only requiring two experiments.

Direct measurement of the work expended between two given points of a circuit.—Let A and B be the potentials at the two ends of a known resistance R, free from self-induction, interposed in the circuit A' and B', the potentials at the two points between which the expenditure of energy is to be measured.

The needle is insulated from the quadrants, and they are connected to A and B. The needle is then connected to A', and the deflection d noted. The general formula of the electrometer gives,

$$d = k\,(A - B)\left(A' - \frac{A+B}{2}\right).$$

The needle is then connected to B' and the deflection d' noted: then,

$$d' = k\,(A - B)\left(B' - \frac{A+B}{2}\right).$$

whence

$$d' = k\,(A - B)\,(A' - B');$$

or

$$d - d' = k\,\text{RCE} = k\text{RW};$$

∴

$$W = \frac{d - d'}{kR} = \frac{a^2}{2R}(d - d').$$

a is determined by *Joubert's* method with a battery of n elements. R is measured in ohms, and the formula gives

$$W = \frac{a^2}{2R}(d - d') \times \frac{1}{9\cdot 81}\text{ kgm. per sec.} = \frac{a^2}{2R}(d - d') \times 1\cdot 35\,\frac{\text{foot-lbs.}}{\text{per sec.}}$$

This method is rigorously accurate, and may be applied either to mere resistance, or to lamps, motors, etc.

ELECTROMOTORS.

Every machine which transforms electrical energy into mechanical work is an *electromotor*. Telegraphs, electric clocks, etc., are in principle electromotors, but this name is only applied to apparatus which effects the transformation continuously so as to produce constant useful mechanical work.

Every mechanical generator of electricity can be used as a motor, as they are reversible; but in practice only continuous currents are used.

Pole reversing motors.—They are more economical than continuous current motors, especially for small power, on account of their simplicity and relative cheapness. There are a great number of different patterns, but in all is to be found a Siemens H armature revolving in a magnetic field produced by a permanent or electromagnet. The best known are those of *Marcel Deprez*, *Trouvé*, *Griscom*, *Cloris Baudet*, *Mercier*, etc. The reversal of polarity of the armature takes place twice in each revolution, so that it is advantageous to reduce the magnetic inertia by reducing the dimensions of the armature as much as possible; for this reason, as M. Marcel Deprez has pointed out, the efficiency of motors with small armatures is far higher than that of the obsolete machines of Froment, Page, Larmenjeat, etc., in which the reversal or successive magnetisation and demagnetisation went on in large magnetic masses. In some recent types there is no iron in the movable or fixed parts in which the current is reversed. As yet we have no information as to results obtained from these little motors, amongst which are those of *Bürgin* (1881), and *Jablochkoff* (1882).

Continuous current motors.—When any considerable power has to be produced continuous current machines are used. Gramme and Siemens machines are most used, their details, and the thickness of wire with which they are wound being modified to suit the currents with which they are to be fed.

Electrical energy absorbed by a motor.—W_e is proportional to the product of the strength of the current passing through it by the difference of potential at its terminals:

$$W_e = \frac{CE}{9\cdot 81} \text{ kgm. per second,}$$

$$\text{or } W = \frac{CE}{1\cdot 35} \text{ foot-pounds per second,}$$

$$\text{or, } W_e = \frac{CE}{746} \text{ h.p. per second.}$$

Electrical work produced by a motor is proportional to the product of the strength of the current passing through it by the back electromotive force E' produced by the motor.

GRAMME MACHINE.

Loss by heating of the wires due to the current is equal to the difference between the electrical energy which it absorbs and the electrical work which it does. When the internal resistance of the motor, and the strength of the current passing through it are known,

$$\text{Heating due to current passing} = \frac{C^2R}{9\cdot 81} \text{ kgm. per second,}$$

$$= \frac{CE}{1\cdot 35} \text{ foot-pounds per second,}$$

$$= \frac{C^2R}{746} \text{ h.p. per second.}$$

Mechanical work given out by a motor is always less than the electrical work.

Deprez motor (1879).—A Siemens H armature revolving between the arms of a ∪-shaped permanent magnet.

D'Arsonval's tests.—Armature 35 mm. long, 30 mm. diameter; wire 1 mm. in diameter; weight of magnet, 1,700 grammes.

Number of Bunsen Cells	4	5	6
Revolutions per minute	140	205	—
Work per minute in kgms.	35	51	60
Current strength C	4·1	4·41	5
Difference of potential at terminals E	79	135	180
Work per gramme of zinc in kgms.	107	134	100

The combustion of one gramme of zinc in the battery produces 1·2 calories (kg.-d.), or 510 kgms. The motor transforms about 26 per cent. of the heat energy produced by the chemical actions in the battery, into mechanical work.

Trouvé's motor.—Siemens armature, with slightly excentric faces; field magnet in the main circuit.

One armature type.—Weight 3,300 grammes. Work at the break 3·75 kgm. C = 20 ampères, with 6 bichromate cells in series.

With six Bunsen cells in series the motor uses 24 grammes of zinc per hour per cell, or 144 grammes for the whole battery. The work produced is 13,500 kgms. per hour.

Work produced per gramme of zinc consumed, 93 kgms.

Gramme machine with permanent magnets.—Laboratory type (*d'Arsonval*). The original magnet, being too weak, was replaced by an Alliance compound magnet. The following are the results of the two most characteristic experiments:

Number of Bunsen Cells	4	6
Work per minute in kgms.	60	100
Strength of current C	3·3	3·4
Difference of potential at terminals E.	4·95	7·5
Electrical energy put into the motor per second	1·63	2·59
Efficiency of motor	·61	·64
Work produced per gramme of zinc consumed	225	250

Fifty per cent. of the heat energy developed in the battery appears as mechanical work. A gramme machine of larger dimensions with 8 Bunsen cells, 12 volts at the terminals, and a current of 1·72 ampères, gave 92 kgms. of useful work per minute, and as much as 368 kgms. per gramme of zinc consumed, or 73 per cent. of the total heat energy. The motor transformed 75 per cent. of the energy actually put into it at the terminals into mechanical work.

Motor worked by a battery.—Practical method of setting up a motor and battery so as to get the maximum work. Fix the motor so that it cannot turn, measure the strength of the current. Let the motor run, and increase its speed until the current is reduced to half the original strength; that speed gives the maximum work, and the efficiency (theoretical) is 50 per cent.; if the motor goes faster the efficiency becomes greater, but the work produced in unit of time becomes less. For the same quantity of work produced per second at different speeds the highest speed gives the best electrical efficiency. This is generally true within certain limits, but in many motors, and probably in all after a certain speed, the loss due to a complex group of phenomena, which is generally called "magnetic friction" (which includes Foucault currents, self-induction, and other phenomena) tends to diminish the efficiency.

Gramme and Siemens motors.—The efficiency of electromotors diminishes with the quantity of work which they are caused to produce per unit of time. This result was obtained by experiments made at Grenoble, in 1883, by a Commission appointed to examine Marcel Deprez apparatus. A glance at the table below at once shows that these motors were working with current strengths and electromotive forces far below those which correspond to their normal use.

Five machines (two Gramme and three Siemens) worked at a constant pressure of 39 volts under the following conditions:

DETAILS.	GRAMME MACHINES.		SIEMENS MACHINES.		
	1	2	1	2	3
Internal resistance in ohms	1·25	1·09	·622	1·307	·615
Current strength in ampères	9·7	11·3	18	19	19·3
Electrical energy put in in kgm. per sec.	38	44	70	74	75·5
Mechanical work taken out	18	19·5	40	39	35·3
Efficiency	·47	·44	·57	·53	·5

Ayrton and Perry's motors (1883).—Consist of a Siemens armature, acting as a field magnet (*i.e.* having the current in it never reversed), rotating inside a Paccinotti ring. The small type, intended to give about 3 horse-power (22 to 25 kilogrammètres per second, or 165 foot-pounds per second), is made in three types of about the same weight, 35 pounds or 16 kilogrammes, to work with differences of potential at the terminals of from 25·5 to 100 volts. The following are the details of some tests:

	25-VOLT TYPE.	50-VOLT TYPE.	100-VOLT TYPE.
Revolutions per minute	1,800	2,000	2,100
Difference of potential at terminals in volts	23	48	98
Current in ampères	25	14·2	6·1
Available work in horse-power	·3	·33	·35
Efficiency	·39	·36	·38

ELECTRICAL TRANSMISSION OF ENERGY.

Based on the principle of reversibility. The current produced by the *generator* works the *receiver* or *motor*.

APPLICATIONS OF ELECTRICITY.

Theoretical case.—Let

E be the electromotive force of the generator (volts);
E' the back electromotive force of the motor (volts);
C the strength of the current (ampères);
R the internal resistance of the generator (ohms);
R' the internal resistance of the motor (ohms);
R'' the resistance of the line (ohms);

When the insulation of the line is perfect,

$$C = \frac{E - E'}{R + R' + R''};$$

let W be the work expended by the generator (transformed into electrical energy),

$$W = \frac{CE}{9 \cdot 81} \text{ kgms. per second} = \frac{CE}{746} \text{ h.p.}$$

$$= \frac{CE}{1 \cdot 35} \text{ foot-pounds per second,}$$

Let W' be the electrical work produced by the motor,

$$W' = \frac{CE'}{9 \cdot 81} \text{ kgms. per second} = \frac{CE'}{746} \text{ h.p.}$$

$$= \frac{CE'}{1 \cdot 35} \text{ foot-pounds per second,}$$

Let W'' be the electrical energy expended in heating the circuit (machines and line):

$$W'' = \frac{C^2 (R + R' + R'')}{9 \cdot 81} \text{ kgms. per second} = \frac{C^2 (R + R' + R'')}{746} \text{ h.p.}$$

$$= \frac{C^2 (R + R' + R'')}{1 \cdot 35} \text{ foot-pounds per second.}$$

$$\textit{Electrical efficiency} = \frac{W'}{W} = \frac{E'}{E}.$$

Or the efficiency is independent of the resistance, and hence of the distance to which the energy is transmitted when the line is perfectly insulated.

Maximum work of the motor.—The work produced being proportional to C E′ follows the variations of these two factors. Other things being the same, it diminishes when R″ increases, that is to say, it diminishes as the distance increases. The distance affects the work produced, but not the efficiency. The work produced is a maximum when

$$E' = \frac{E}{2}.$$

The electrical efficiency being 50 per cent.

As E′ increases, the efficiency increases from ·5 to 1, but the work produced diminishes; in the limit when E′ = E the efficiency becomes 1 or 100 per cent., and the work produced = 0.

As the work produced diminishes as the whole resistance of the circuit increases, in order to transmit the same quantity of work at the same theoretical efficacy, it is necessary to increase the e. m. f. both of the generator and motor, and to diminish the current as the distance increases. The three equations which give W, W′, and W″ enable the theoretical values of C, E, and E′ to be calculated, when the work to be transmitted, the work to be expended, and the loss which can be allowed by heating of the circuit are known, the resistance of the line being determined by the distance and by practical considerations of the cost of erection.

Theoretical limit of the work transmitted by a line of given resistance.

—Theoretically an infinite quantity of work can be transmitted by a line of any length by using sufficiently high electromotive forces. In practice a limit is soon reached, on account of the danger of high electromotive forces and the loss by leakage, which increases very rapidly as the electromotive force goes up. If, for example, we have the conditions that the e. m. f. of the generator is not to exceed 3,000 volts, and the loss by heating of the line to be 20 per cent. of the work expended, it is easy to calculate the theoretical limit of the work transmitted, the current strength, and the work recovered by taking, say, a final efficiency of 50 per cent.; the results are given in the following table:

Theoretical limit of transmission of power.—(E = 3,000 volts, heating of the line, 20 per cent. of the work expended by the generator. Loss by leakage = 0.)

Resistance of the line in ohms.	Maximum strength of the current in ampères.	Maximum work transformed into electrical energy by the generator in horse-power.	Work recovered at the motor in horse-power for an efficiency of 50 per cent. in horse-power.	Loss of energy by resistance of the line in horse-power.
5	120	473·3	236·6	94·7
10	60	236·6	118·3	47·3
20	30	118·3	59·1	23·6
50	12	47·3	23·6	9·5
100	6	23·6	11·8	4·75
200	3	11·8	5·9	2·37
500	1·2	4·7	2·36	·95
1,000	·6	2·35	1·18	·47

Practical case.—In practice the work put in is always greater, and the work taken out less, than the values given by the formulæ, on account of friction, secondary actions, leakage losses, etc.; so the formulæ will have eventually to be modified by affecting them by practical coefficients, which are as yet for the most part unknown. It is by no means surprising that direct measurements differ widely from the results given by calculation.

Efficiency.—There is no word with such divers meanings as *efficiency* as applied unqualified to electrical work. Its exact mechanical meaning is: *The ratio of the work taken out of a machine to the work put into it.* But in electrical nomenclature, on account of the number and variety of effects which a current can produce, it has many meanings. To avoid confusion, the word efficiency ought always to be accompanied by some name or adjective which exactly defines its particular meaning in the place where it is used.

Electrical efficiency of a generator.—The ratio of the total electrical energy given out to the mechanical work put in; that is to say, between the whole work converted into electrical energy and the whole work put in.

Available electrical energy.—Ratio of the electrical energy at the terminals available in the external circuit to the whole mechanical work put in.

Mechanical efficiency of a motor.—Ratio of the mechanical work taken out, measured at the break, to the electrical work at the terminals.

Electrical efficiency of a system of transmission of power.—Ratio of the electrical work transformed into mechanical work by the motor to the mechanical work transformed into electrical work by the generator. When there is no loss by leakage, the electrical efficiency is $\frac{E'}{E}$.

Mechanical or commercial efficiency of a system.—Ratio of the mechanical work done by the motor, measured at the break, to the mechanical work put into the generator measured at the dynamometer. This kind of efficiency is what ought to be meant when efficiency is spoken of without any qualifying epithet.

Increase of resistance of the armature due to self-induction.— Applicable both to motors and dynamos (*Ayrton and Perry*). If L be the coefficient of self-induction of the coil, C the current passing, and n the number of revolutions per minute, then the loss of energy per second is

$$\frac{\pi n L C^2}{120},$$

which corresponds to an increase in the resistance of the armature equal to

$$\frac{\pi n L}{120} \text{ ohms.}$$

To find the coefficient of self-induction of a coil.—The simplest method is that adopted by Lord Rayleigh for correcting for self-induction in his determination of the ohm by the British Association method.

Arrange the coil in a Wheatstone's bridge as if to measure its resistance, using a very delicate undamped galvanometer. Get a perfect balance, and call the resistance of the coil thus found P. Now close the battery key before closing the galvanometer key; on the galvanometer key being pressed down the needle will be momentarily deflected; observe the throw, call this α. Now insert an additional resistance δP in the same circuit as the coil. Close the battery key before the galvanometer key (in the usual way), and observe the throw of the needle; call this β. Also observe the time T of one-half complete vibration of the needle. Then if L be the coefficient of self-induction,

$$\frac{L}{P} = \frac{\delta P}{P} \frac{T}{\pi} \frac{2 \sin \frac{1}{2}\alpha}{\tan \frac{1}{2}\beta}.$$

Marcel Deprez's experiments *between Miesbach and Munich* (1882). — Two A gramme machines wound with fine wire connected by a double overhead telegraph line of iron wire 4·5 mm. in diameter. Diameter of wire on machines ·4 mm.

Resistance of line	950·2 ohms.
,, generator	453·1 ,,
,, motor	453·4 ,,
Strength of current at generator	·519 ampère.
Difference of potential at terminals of motor	850 volts.
Available electrical work at motor	·426 h.p.
Electrical work of generator (calculated)	1·01 ,,
Effective work taken out of motor	·245 ,,
Electrical efficiency (calculated)	38·9 per cent.
Mechanical efficiency (calculated)	22·1 ,,

Experiments at the Chemin de Fer du Nord (1883).—A Marcel-Deprez machine, with double gramme ring as generator and a D gramme as motor. Diameter of wire, 1 mm. (The machines being arranged side by side made leakage on the line an advantage, instead of a disadvantage as it is when the machines are at opposite ends of a double line.) The line was made of telegraph wire 4 mm. in diameter, 17 kilomètres long, and of 160 ohms resistance.

Resistance of generator	56 ohms.
,, motor	83 ,,

The following are the results of two series of tests made by Messrs. Tresca, Hopkinson, and Cornu:

	1st series.	2nd series.
Strength of current in ampères	2·559	2·687
Generator.		
Power expended at the dynamometer	6·21	10·4
Number of revolutions per minute	590	814
Difference of potential at terminals	1290	1865
Electrical energy produced	4·42	6·81
Motor.		
Number of revolutions per minute	365·8	595
Difference of potential at terminals	908	1485
Electrical energy put into the motor	3·12	5·42
Mechanical work (measured at the break)	·326	·317

Between Vizelle and Grenoble (1883). — Generator, a Gramme machine with two rings coupled in series (Marcel Deprez type, No. 10) on the same axis. Field magnets, two U-shaped electro-magnets.

Motor, D Gramme machine altered for the experiment.

Distance of transmission 14 kilomètres, overhead line formed of two wires (lead and return) of silicium-bronze two millimètres in diameter.

TRANSMISSION OF ENERGY.

Resistance of the line . . . . . . 167 ohms.
Resistance of generator :
    Field magnets . . . . . . 20·1 ⎫
    Rings 2 × 18·3 . . . . . . 36·6 ⎬ 56·7 ohms.
Resistance of motor :
    Field magnets . . . . . . 61 ⎫
    Rings . . . . 36 ⎬ 97

DETAILS.	EXPERIMENTS.	
	D 2.	N 1.
Revolutions per minute of generator . . .	995	1140
,, ,, motor . . .	618	875
Motive power at axis of break in chevaux vapeur	16·28	16·9
Motive power (gross)	12·61	11·56
Work with transmission deducted . .	12·27	11·18
Work received at the break	6·33	6·97
Mechanical efficiency	51·6	62·3
Mean current strength in ampères . .	3·46	2·35
Electromotive force of generator E . .	2848	3146
,, ,, motor e . .	1737	2231
Electrical efficiency $\frac{e}{E}$	60·9	70·8

Loss by leakage on the line.—Experiments made by means of nitrate of silver voltmeter on the silicium bronze line when recently put up in a temporary way, gave the following results :

DETAILS.	EXPERIMENTS.	
	1	2
Electromotive force of the generator (in volts)	2808	3128
Difference of potential at terminals ,,	2627	2934
Strength of current at Vizelle in ampères . .	3·268	3·514
,, ,, Grenoble ,, . .	3·099	3·282
Loss per cent.	5·1	6·6

GENERAL USEFUL FORMULÆ.

Let
    E be the e. m. f. of the dynamo.
    e be the back e. m. f. of motor.
    R be the resistance of dynamo.
    R' be the resistance of line.
    r the resistance of motor.
    W, mechanical work put in to the dynamo.
    w, work taken out of motor.

R

To test, put an ammeter in circuit and observe the current passing C, place a voltmeter between the terminals of the motor, and observe the difference of potential, P; we then have

$$e = P - Cr \qquad E = P + C(R + R').$$

$$C = \frac{E - e}{R + R' + r}.$$

Now for motor, electrical work put in is

$$CP,$$

and whole convertible or available electrical work is

$$Ce, \text{ or } C(P - Cr).$$

Maximum possible efficiency of motor under the conditions of the test $\quad = \dfrac{P - Cr}{P}$ or $\dfrac{e}{P}$.

Real efficiency $\quad = \dfrac{w}{CP}$.

Ratio of real efficiency to maximum possible efficiency $\quad = \dfrac{w}{C(P - Cr)}$.

Loss due to friction, self-induction, and the group of phenomena called magnetic friction $\quad = C(P - Cr) - w$.

A test should be made to ascertain the loss by mechanical friction; call this L,

Then the loss due to electrical and magnetic actions $\quad = C(P - Cr) - (L + w)$.

If this loss be high it shows that the electrical design of the motor is faulty; if L be high it shows that the mechanical design is faulty.

For any system of transmission of energy:

The maximum possible efficiency $\quad = \dfrac{e}{E}$ or $\dfrac{P - Cr}{P + C(R + R')}$.

The real electrical efficiency $\quad = \dfrac{w}{C\{P + C(R + R')\}}$.

The real mechanical efficiency $\quad = \dfrac{w}{W}$.

Percentage of maximum possible efficiency to efficiency obtained . . . . . $= 100 \dfrac{w\{P + e(R + R')\}}{W(P - Cr)}$

Loss by heating in dynamo . . . . $= C^2 R$.
Loss by heating in line . . . . $= C^2 R'$.
Loss by heating in motor . . . . $= C^2 r$.

Thus, in any system, there is to be considered as to efficiency the efficiency of the dynamo and the efficiency of the motor; and as to quantity of power which this system will transmit, the whole resistance of the circuit. Having a good type of dynamo and motor, in order to get high efficiency and plenty of power, E and e should be high, so that $\dfrac{e}{E}$ may be near unity and $E - e$ may be large. The resistance of the whole system should be low; but as a practical rule, to avoid expense in erection of the line, the resistance of the system may be allowed to increase as E and e increase. Again, as in order to increase E and e we must increase R and r, R' may increase very rapidly as E and e increase. As yet so little has been done on a large scale in electrical transmission of energy, and the systems having been for the most part experimental, and therefore not been kept running long, that there is a great dearth of practical information on the subject.

According to the purpose for which the motor has to be used, very different types will be selected. Where weight is of no moment, several horse-power can already be obtained with very high efficiency from the Gramme and the Bürgin dynamos used as motors. At Vienna, Siemens Brothers obtained from the Siemens (Heftner-Altneck) continuous current dynamo, used as a motor on board their electric boat, supplied by the Power Storage Company's accumulators (the Sellon-Volkmar-Faure-Swan combination), as much as 6 horse-power with an efficiency (work to CP) of 75 per cent. For light motors one of the best results lately was from some of Ayrton and Perry's motors:

Speed	h.p. given out.	Efficiency (work to CP).	Weight of motor.
2,100 revs. per minute	.35	38%	37 lb.
1,340 ,, ,,	·75	47%	75 ,,
1,600 ,, ,,	2·1	51%	127 ,,
2,000 ,, ,,	2·6	not observed.	127 ,,

This series goes also to show that horse-power per unit of weight and efficiency increase as the weight increases in motors of the same type.

The varying conditions of driving e. m. f., speed, and lead of brushes all have the greatest possible influence, both on the horse-power and efficiency of motors. When any one of these variables is fixed, the other two will each have certain definite values in order to get the best effect. The values also, unfortunately, are not always the same for maximum h.p. and maximum efficiency, though probably the better the design of the motor, the more nearly the arrangements for maximum power will give maximum percentage of $\frac{e}{E}$ efficiency. As yet, the variation of the secondary effects (self-induction and magnetic friction) make the arrangements for the highest power always differ widely from Jacobi's rule of $\frac{e}{E} = \frac{1}{2}$.

ELECTRIC LIGHTING.

Electric light is the direct application of the heat produced by currents. This heat is utilised and concentrated into the smallest possible space, so as to raise the temperature as high as possible, and so cause certain bodies to become luminous. According to the nature of the conductor passed through and made luminous by the current, three large classes of practical means of producing electricity may be formed.

1. *Rarefied air*, made luminous by the passage of a current.
2. *The voltaic arc*, formed by the passage of a current in air raised to a high temperature; this air heats the carbons or other refractory bodies by direct contact and renders them incandescent.
3. *Incandescence.*—Solid matter, generally carbon, raised directly to a high temperature by the passage of a current. We will only discuss here the voltaic arc and incandescence, light produced in vacuum tubes not having as yet been practically applied.

VOLTAIC ARC.

The voltaic arc is almost always produced between two pencils of artificial or agglomerated carbon, and the light is due, not to the arc properly so called, but to the incandescence of the carbons raised to a high temperature by the passage of the current. The combustion of the carbons causes them to wear away, and it becomes necessary, in order that the light may be steady and to prevent its going out, to keep the carbons at a suitable distance from each other, either by hand or by the aid of properly-devised lamps, which cause the carbons to approach each other automatically in different ways, which are more or less perfect, more or less simple, or more or less economical.

Classification principles of arc lamps.— The arc lamp is said to be *monophotal* when its system of governing is such that only one source of light can be worked from each source of electricity, whether battery or machine. It is *polyphotal* when several can be placed on one circuit, either in series or in parallel arc or in several groups, according to the character and power of the machines which feed them. Generally, regulation is based on electro-magnetic actions, and is produced

(1) *By current strength.*—The mechanism tends to keep the strength of the current constant.

(2) *By a shunt circuit.*—The regulating electro-magnet, wound with fine wire, is arranged as a shunt between the terminals of the lamps, and tends to keep the difference of potential between these two points constant.

(3) *By differential action.*—The regulating machinery tends to keep a certain equilibrium between the two factors, strength of current and difference of potential at the terminals, and comes into action as soon as one or other of these two elements tends to become weaker or too strong.

(4) *Various lamps.*—There are a certain number of lamps based on different actions and difficult to classify. In some the carbons are re-adjusted at regular intervals of time, as one minute or half a minute (*Brockie*). Others keep a constant geometrical distance between the carbons, either by the wearing away itself (*Rapieff*), or by arranging carbons side by side (electric candles), or by utilising the heat of the arc to produce the bringing together of the carbons at the desired moment (*Solignac*).

As to mechanical arrangements, they vary infinitely, and the fruitfulness of inventors in this direction is inexhaustible. Gearing, cords, springs, weights, motors, water, mercury, compressed air, electro-magnets of all sorts, solenoids, double-wound magnets, ratchets, mechanical and magnetic breaks have been employed or proposed; and any classification based on these characteristics would be of no scientific or practical interest.

Alternating and continuous currents.—When an arc lamp is fed by alternating currents, the wear of the two carbons is the same if they be horizontal; if they be vertical, the upper carbon wears away a little faster than the lower one, in the ratio of about 108 to 100 for carbons of the same quality and same diameter. With continuous currents, the positive carbon, which is generally placed above, wears away twice as fast as the negative carbon, and is hollowed out into a crater.

When it is desired to keep the light fixed in space, account must be taken of the different rates of wearing according to the nature of the current employed.

Resistance of the voltaic arc.—As yet there is the greatest uncertainty as to the true value of the resistance of the voltaic arc, because as yet the true value of the back e. m. f. developed in the arc by the passage of the current is unknown. We shall give no figures, as opinions are so contradictory; further, such figures are absolutely useless for purposes of calculation, because, if we know the strength of the current and the difference of potential at the terminals of a given arc lamp, we have all that is necessary for calculating its expenditure of electrical energy and the conditions which must be fulfilled by the machine which has to feed it.

Electrical energy absorbed by an electric light.—Let C be the strength of the current in ampères necessary for the good working of an electric lamp, arc, candle, incandescent, etc., and E be the difference of potential in volts at the terminals of the lamp. The electrical energy W absorbed by it will be:

$$W = \frac{CE}{9 \cdot 81} \text{ kgms. per second}, = \frac{CE}{1 \cdot 35} \text{ foot-pounds per second};$$

$$\text{or, } W = \frac{CE}{746} \text{ h.p.}$$

Dividing W by the lighting power of the lamp L in candles or carcels, we get the price of unit of light in kilogrammètres or foot-pounds of electrical energy. Dividing the lighting power L by W, we get the number of units of light which one kilogrammètre or foot-pound of electrical energy can furnish. Either of these two numbers gives the absolute value of a given electric light in relation to the quantity of electrical energy which it consumes. Of course the number of units of light per kilogrammètre of effective work furnished by the electrical generator is always lower than the figures given by the above formula, because it necessarily includes the coefficient of the efficiency of the machine, the loss due to the resistance of the conductor, and other causes of loss independent of the light properly so called.

CARBONS.

When the carbons are too thin they give a very intense light, but burn away too quickly; when the carbons are too thick very deep craters are formed so that the light is obscured and diminished. Coating with metal increases the life of the carbons.

1. Trimming of bare carbons.

2. Trimming of coppered carbons.

3. Trimming of nickel-plated carbons.

Fig. 44.—Experiments of M. E. Reynier on the Trimming of Bare and Plated Carbons.

Bare and plated carbons (*E. Reynier*).—Experiments made in the shops of Sautier and Lemonnier with an A gramme machine and homogeneous Carré carbons from the same sample. Photometric measurements taken by projecting the light forward by means of an arrangement which Mons. Lemonnier considers sufficient for practical purposes.

However, when the sections of carbons and the conditions of their surfaces are varied, this method may not be sufficient. Thus, the photometric values in the table must be considered as mere approximations until less arbitrary measurements have been made.

The metallic plating was not very adherent; it often scaled off. By improving the plating processes, a good adherent metallic deposit can certainly be obtained. On the positive carbon, the ordinary method of trimming is good with copper, and excellent with nickel. On the negative carbon, the trimming which is a little too long for bare carbons appeared to be a little too short with plated carbons.

Further, the metal sometimes remains round the carbon at the cut part, forming an injurious projection. This inconvenience might, no doubt, be avoided by plating the negative carbon with brass.

DIMENSIONS.	STATE OF THE SURFACE.	LENGTH CONSUMED IN ONE HOUR.			LENGTH OF TRIMMING.		PHOTOMETER.
		Positive.	Negative.	Total.	Positive.	Negative.	
		mm.	mm.	mm.	mm.	mm.	
$d = 7$ mm. $s = \cdot 3846$ cq.	Bare	166	68	234	53	23	947
	Blackened copper	146	40	186	24	10	?
	Blackened nickel	106	38	144	12	7	947
$d = 9$ mm. $s = \cdot 6358$ cq.	Bare	104	50	154	45	22	528
	Blackened copper	98	34	132	27	7	553
	Blackened nickel	68	36	104	21	7·5	516

Remarks.—Independently of the improvement in the cutting of the positive carbon, the use of nickel increased the time of consumption 50 per cent. in the case of 9 mm. carbon, and 62 per cent. in that of 7 mm. carbon. Coppered carbon lasts for a length of time between that of bare carbon and nickeled carbon. For equal sections, the metallic plating does not seem to affect the luminous efficiency; but the sectional area appears to have a very sensible effect on the lighting power. Thus, 7 mm. carbons have given much higher readings on the photometer than 9 mm. carbons.

GRAMME MACHINES AND PROJECTORS USED IN THE FRENCH NAVY.
(Sautter and Lemonnier).

	M.	AG.	CT.	CQ.	DQ.
Nominal power in carcels	200	600	1600	2500	4000
Speed in revolutions per minute	1600	820	675	1380	495
Mean length of arc in mm.	3	4	4	4·5	6
Strength of current in ampères	13·5	24·5	48	65	70
Diameter of carbons in mm.	9	13	18	18	20
Number of carcels, mean	226	490	1015	1241	2198
Number of carcels, lamp inclined (utilised in the projector)	625	1200	2500	3300	6000
Work absorbed in h.p.	·93	2·69	5·14	7·94	9·99
Weight of machine in kilogrammes	73	185	390	390	1000

The M type is used on steam launches. The AG type on despatch boats. The CT type on ironclads. The CQ and DQ types for coast defence.

Abdank-Abakanowicz lamp (1882).—Differential for continuous currents. Bare Siemens carbons. Positive 12 mm.; negative 10 mm. Normal current: 10·5 ampères. Difference of potential at terminals 42 to 44 volts.

Gülcher lamp (1881).—A mean current of 15 ampères gives a light of 113 carcels in the horizontal plane, and 136 carcels at an angle of 34°. The positive carbon being above.

Tests at the Paris Electrical Exhibition (1881).—The Commission was formed by Messrs. *Allard, Joubert, F. Le Blanc, Potier,* and *Tresca*. They used certain terms to express the quantities which they measured, which we will explain and define in order that the tables may be readily understood.

Electric horse-power, arc horse-power (cheval electrique, cheval d'arc). —Electrical energy produced or consumed by a machine calculated from the e. m. fs., resistances, and current strengths expressed (by the Commission) in chevaux-vapeur of 75 kgm. per second. In the tables we will give these values in horse-power of 33,000 foot-pounds per minute.

Total mechanical efficiency (rendement total mécanique).—Ratio of effective work obtained to total mechanical work, deducting that which is used in mechanical transmission.

Mechanical efficiency of arcs (rendement mécanique des arcs).—Ratio of the effective work done in the arcs to the electrical energy put into the lamps.

Electrical efficiency of the arcs (rendement electrique des arcs).—Ratio of the electrical work in the arcs to the total electrical work.

The results obtained by this Commission will be found in the tables on pages 266, 267.

ELECTRIC CANDLES.

The first candle was invented by *Paul Jablochkoff* in 1876. It consists of two pencils of carbon separated by a layer of insulating material. In order that both carbons may wear away at the same rate alternating currents are employed. A form has been tried with the positive carbon thicker than the negative so as to use continuous currents. It has not come into practical use. Other inventors, *Wilde* (1879), *Jamin* (1879), *Debrun* (1880), have brought out candles without an insulating layer between the carbons with automatic relighting arrangements, but they have not been much used. Almost the only electric candle in use at the present day is Jablochkoff's.

TABLE OF TESTS OF CONTINUOUS

DETAILS.	FORMULÆ.	I GRAMME. 1 lamp.	II. JURGENSEN. 1 lamp.	III. MAXIM. 1 lamp.	IV. SIEMENS. 1 lamp.
Speed of generator	T. pr min.	475	800	1017	737
Effective motive power	W. H.P.	16·13	21·68	4·07	4·44
Res. of machine in ohms	—	·33	·45	·7	·66
Res. of circuit, neglecting lamps	—	·10	·82	·25	·12
Total res.	R ohms	·43	1·27	·95	·78
Strength of current in ampères	C amp.	109·2	90	33	35
Fall of potential at lamp in volts	E volts.	53	58	53	53
Work in whole circuit	$\frac{RC^2}{75\,g}$	6·97	13·99	1·41	1·29
Work in one lamp	$\frac{EC}{75\,g}$	7·87	7·09	2·37	2·52
Work in lamps	w H.P.	7·87	6·97	2·31	2·52
Total electrical work	T'	14·84	20·96	3·72	3·81
Mean e. m. f.	$nE + RC$	102	172	84	80
Diameter of carbons	mm.	20	23		18
Lighting power horizontal	carcels	952	607	246	210
,, ,, maximum	carcels	1960	—	465	805
,, mean spherical ,,	l	966	688	239	306
,, total mean spherical ,,	$L = nl$	966	688	239	306
Total mechanical efficiency	$\frac{W'}{W}$	·92	·97	·91	·86
Mechanical efficiency of the arcs	$\frac{w}{W}$	·43	·32	·57	·57
Electrical efficiency of the arcs	$\frac{w}{W'}$	·53	·33	·62	·66
Carcels per mechanical h.p.	$\frac{L}{W}$	60	31·7	58·7	68·9
,, ,, electrical h.p.	$\frac{L}{W'}$	65·1	32·8	64·2	80·3
,, ,, arc h.p.	$\frac{L}{w}$	128·8	98·7	103·5	121·4
,, per ampère	$\frac{l}{C}$	8·85	7·64	7·24	8·74

TESTS OF MACHINES AND LAMPS.

CURRENT MACHINES AND LAMPS.

V. SIEMENS. 2 lamps.	VI. BURGIN. 3 lamps.	VII. GRAMME. 3 lamps.	VIII. GRAMME. 5 lamps.	IX. SIEMENS. 5 lamps.	X. WESTON. 10 lamps.	XI. BRUSH. 16 lamps.	XII. BRUSH. 40 lamps.	XIII. BRUSH. 38 lamps.
1330	1535	1695	1496	826	1003	770	700	705
5·31	5·32	8·11	8	5·05	130·1	13·39	29·96	33·35
1·68	2·8	0·52	4·57	7·05	1·88	10·55	22·38	22·38
·13	1·5	1·25	0·62	4·5	1·5	2·56	2·6	7·9
1·81	4·3	1·77	5·19	11·55	3·38	13·11	24·98	30·28
26·2	18·5	19	15·3	10	23	10	9·5	9·5
44·5	41	53	40·8	47·4	32	44·3	44·3	44·3
1·69	2	0·87	1·65	1·57	2·43	1·79	3·07	3·72
1·59	1·02	1·369	1·04	0·64	1	·6]	·57	·57
3·18	3·08	4·11	5·2	3·2	10	9·6	21·88	20·79
4·87	5·08	4·98	6·85	4·77	12·43	11·39	24·95	24·51
136	203	193	328	353	398	840	2009	1971
14	13	14	12	10	9 and 10	11	11	11
142	50	155	112	67	92	37	63	63
537	227	357	184	72	154	76	78	78
205	82	167	102	52	85	38	39	39
410	243	501	510	260	850	608	1560	1482
·92	·95	·62	·86	·94	·95	·85	·83	·73
·6	·58	·51	·65	·63	·77	·72	·73	·62
·65	·61	·83	·76	·67	·8	·84	·87	·85
77·2	46·2	61·8	63·8	51·5	65·3	45·4	52·1	44·4
84·2	48·4	100·4	74·5	54·6	68·4	53·4	62·6	60·5
129·3	79·9	121·6	93·1	81·3	85	63·3	71·7	71·4
7·82	4·43	8·79	6·67	5·2	3·7	3·3	4·11	4·11

Jablochkoff candles.—4 mm. candles (*J. Joubert*).

Strength of current	8 to 9 ampères.
Fall of potential at base of candle . .	42 to 43 volts.
Electric energy expended	{ 34 to 39 kgms. per second, ·45 to ·51 h. p.

In practice about one indicated horse-power per light is allowed.

The light in front of the candle being taken as unity, that at the side is ·57.

The mean intensity of the light is ·9 of the maximum intensity.

Experiments made with four candles burning at once, two facing and two turned sideways, gave the following results:

Bare light	37·5 carcels per candle.
Intensity of light facing . . .	45 ,, ,,
Mean intensity	41 ,, ,,
With ordinary opal globe . . .	22·5 carcels, or ·575 of the mean intensity.
With clear globe	27·5 to 30 carcels, or ·67 to ·75 of the mean intensity.

ELECTRIC CANDLES.
Tests of the Commission at the Paris Exhibition, 1881.

DETAILS.	Derun Candle.	Jablochkoff Candle. Gramme machine	Jablochkoff Candle. Méritens' machine.	Jamin Candle. — 32	Jamin Candle. — 48	Jamin Candle. — 60
Mean spherical intensity of light	27·4	20·2	23·7	16	17·4	9·4
Strength of current in ampères .	10	7·5	8·5	6·1	5·1	3·5
Difference of potential in volts .	50	43	42	77	69	74
Electrical energy absorbed in kgms.	65	32·5	32·4	47	35·7	25·3
Carcels per arc horse-power .	32	46·9	52·3	25·6	36·9	27·6
Carcels per ampère. . . .	2·74	2·69	2·79	2·69	3·41	2·69

Lampe-soleil, or sun-lamp (1881).—Alternating currents. The same means have been tried with this lamp as with the Jablochkoff candle to make it use continuous currents, with the same results. Incandescence and arc, may be considered as forming a link between electric candles and semi-incandescent lights. Yellow light thrown in one direction. Consumption of carbon very slow, being from 7 to 10 mm. per hour per carbon. Carbons segmental in section. Seventy-carcel lights (in the direction of maximum intensity) require currents of from 4 to 6 ampères, the 600-carcel lamps work with 23 ampères. The difference of potential at the terminals was not given by the experimenters, Messrs. *Bède*, *Desguin*, *Dumont*, *Rousseau*, and *Wauters*, but the experiments show that the light and the current diminish rapidly by more than 50 per cent. after the lamp has been burning for one hour.

INCANDESCENCE.

Incandescent light is produced by the heating of some refractory body by the passage of a current. Platinum and iridium have been tried in incandescent lamps, but as yet without much success. The capital fault of these substances is that if the temperature be low the light is feeble, and the efficiency low; if the temperature be raised sufficiently to get good conditions the slightest accidental increase of current strength melts the incandescent body and extinguishes the lamp. The only practical application of platinum is in the *polyscope* of M. *G. Trouvé*. Carbon is now exclusively used in incandescent lamps.

Carbon incandescent lamps may be divided into two classes :

1st. *Semi-incandescent lamps*, in which the incandescent carbons are exposed to the air, and are slowly consumed, being automatically fed forward. In most of these lamps small voltaic arcs are formed.

2nd. *True incandescent lamps enclosed in an air-tight globe*, in which a filament of carbon, protected from the chemical action of the air by the exhaustion of the globe, is raised to incandescence by the passage of a current, and produces light for a very long time before it is destroyed. The length of time the filament lasts, or the *life of the lamp*, is often more than 1,000 hours of total lighting; some good makers are now enabled to guarantee an average life of 800 hours.

SEMI-INCANDESCENT LAMPS.

Reynier lamp (1878).—Carré carbon 2·5 mm. in diameter.

Length of incandescent part	13 mm.
Strength of current	27 ampères.
Resistance of lamp	·2 ohm.
Difference of potential at terminals	5·4 volts.

Electrical energy expended . . . . . 14·86 kgm. per second,
or 107·5 foot-pounds per sec.
Intensity of light in a zone 15° above the horizontal . . . . 12 carcels.

Werdermann lamp (1880) (*Cabanellas*). — Carbon 4·5 mm. in diameter.

Strength of current . . . . . . 50·5 ampères.
Difference of potential at terminals . . . 6·75 volts.
Electrical energy expended . . . . . 34 kgm per second
or 245·9 foot-pounds per sec.
Light (mean horizontal) . . . . . 34 carcels.
Energy expended per carcel . . . . . 1 kgm. per second,
or 7·23 foot-pounds per sec.
Number of carcels per h.p. of electrical energy 74·25.

PURE INCANDESCENT LAMPS.

There are only four or five out of the immense number of proposed incandescent lamps which are now being commercially used.

Edison lamp.—A filament of carbonised bamboo bent into a U shape, and enclosed in a glass globe exhausted to about one-millionth of an atmosphere.

Swan lamp.—Filament of cotton parchmentised by immersion in sulphuric acid, carbonised, twisted into a sort of curl, and enclosed in an exhausted globe.

Maxim lamp.—Filament of carbonised Bristol card-board cut out in the form of an M enclosed in a globe containing a rarefied atmosphere of benzoline.

Lane-Fox lamp.—Filament of "chiendent" or bass, carbonised and enclosed in an exhausted globe.

All these forms depend on one method of preparing the carbons, which is as follows: After the filament is fixed in the lamp the globe is exhausted to about one-millionth of an atmosphere; a current is then sent through the carbon filament whilst the exhausting pumps continue to work, the current is gradually increased until the filament is brought up to a bright white heat; it is then kept steady, and the pumping continued for some time. In most lamps the globe is then sealed; but in the Maxim and some of Lane-Fox's, etc., some hydrocarbon vapour is now introduced (the filament being kept hot) and pumped out again, this process being repeated several times before the globe is sealed; the object being to deposit carbon in the pores of the filament.

INCANDESCENT LAMPS.

In other respects these lamps resemble each other very closely in mode of manufacture, except in the case of the Lane-Fox lamp, in which, instead of the carbon filament being fixed directly to the platinum wires sealed through the glass, there is a somewhat complex arrangement of platinum wire, mercury cups, copper wire, plaster of Paris, and cotton wool, the object being to reduce the expense of long lengths of platinum wire.

Otherwise, the difference is mainly in the substance from which the carbon filament is prepared. Mr. Edison and his followers prefer a substance which retains a fibrous structure after carbonising, whilst Mr. Swan and his school prefer to destroy all structure in the material as far as possible so as to obtain an homogeneous carbon. Mr. Crooks, besides modifying the glass envelope with a view of cheapening the manufacture, uses a process of destroying structure which he claims as more complete than Mr. Swan's parchment paper or sulphuric acid process. Mr. Crook's lamp has been but little seen in public, and it would not be possible at present to express an opinion on the relative merits of the different forms of lamps. The following tables give the results of some tests made on some of the forms:

EDISON LAMPS.

DETAILS.	1880 TYPE.*	1882 TYPES.† A	B
Resistance in ohms hot	74·50	140	70
Strength of current in ampères	1·08	0·75	0·75
Difference of potential at terminals	80·50	100	50
Candle power	10	16	8
Electrical energy expended in kgm. per second	8·27	7·5	3·8
Electrical energy expended in foot-pounds per second	59·72	54·26	27·48
Number of lamps per electrical h.p.	9	10	20
Number of candles per h.p.	119	158·4	158·4

* *Henry Morton* (Tests made at the Stevens Institute, Hoboken).
† *R. V. Picou* (mean figures).

In practice about one effective horse-power put into the generating dynamo is allowed for 8 A lamps or 16 B lamps (1882 pattern).

INCANDESCENT LAMPS.

(Tests of the Commission of the Paris Electrical Exhibition, 1881.)

DETAILS.	MAXIM.	EDISON.	LANE-FOX.	SWAN.
Resistance in ohms hot	43	130	28	31
Strength of current in ampères	1·74	0·70	1·77	1·55
Difference of potential at terminals, in volts	75	91	50	48
Electrical energy absorbed in kgms. per second	13·23	6·50	8·95	7·62
Electrical energy absorbed in foot-pounds per sec.	95·55	47	65	55·11
Mean spherical carcel power	2·80	1·57	1·64	2·19
Carcels per mechanical h.p.	16	18·35	13·95	21·85

Siemens and Halske's incandescent lamps (1883).—All the lamps made by the firm of Siemens and Halske work with a difference of potential at the terminals of 105 volts. They are constructed of three different patterns. To facilitate comparison we have contrasted them with Edison A lamps.

DETAILS.	NEW LAMPS BY SIEMENS AND HALSKE.			EDISON LAMP.
	PATTERN II.	PATTERN IV.	PATTERN VI.	PATTERN A.
Normal candles	12	16	25	16
Volts	100	100	100	100·4
Ampères	·41	·55	·8	·71
Ohms (hot)	244	182	125	141
Watts	40·5	55	80	71
Normal candles per electrical h.p. in the lamp	213	208·9	224	161·2

High resistance lamps.—In order to lessen the loss by heating of the main conductors, and in order to increase the number of lamps arranged in parallel arc on a given lighting system, and yet to prevent the lamps from varying in brilliancy too much as more or fewer are turned on, high resistance lamps are now made, of which the following is an example.

INCANDESCENT LAMPS. 273

Swan lamp (1883).

Resistance cold	292 ohms.
,, hot	150·8 ,,
Strength of current	·63 ampère.
Difference of potential at terminals	95 volts.
Candle power	20·9 candles.
Energy absorbed per lamp	7 kgms. per second, = 50·63 foot-pounds per second.
Number of lamps per electrical h.p.	10·9.
Number of candles per electrical h.p.	228·6.

The Nothomb lamp (1883).—Filament of cellulose carbonised in a carburetted atmosphere; 1 mm. broad and ·4 mm. thick. The A, B, and C patterns work with currents from 1 to 3 ampères, and e. m. f. of from 45 to 100 volts, giving respectively 30, 50, and 100 candle-power. The D pattern of 300 candle-power has three filaments. When they are coupled in series the lamp requires 300 volts and 3 ampères. When they are coupled parallel it requires 100 volts and 9 ampères.

Boston or Bernstein lamp (1883).—The filament is formed of a thin-walled tube of braided silk, carbonised on a bed of graphite. The pattern called the 50-candle lamp gives normally 60 German candles. It works with a current of 5·4 ampères, and a difference of potential at the terminals of 28 volts. The expenditure of energy is 15 kilogrammètres per second (123 foot-pounds per second), or 303 candles per electrical horse-power.

The 90-candle pattern requires 8·5 ampères and 34 volts at the terminals, or 29 kgm. per second (= 209·76 foot-pounds per second). The same lamp has been pushed without destroying the filament up to 11·8 ampères and 46 volts at the terminals, thus absorbing 54 kgm. per second (= 380·6 foot-pounds per second), and giving 460 candle-power.

Small lamps.—The small lamps used for ladies' head-dresses, scarf pins, etc., are of very small dimensions, and require remarkably little energy to produce a quantity of light, which it is difficult to measure exactly, but which may be estimated as about equal to one good composite candle. The smallest size (12 mm. long and 6 mm. in diameter, will work with 3·7 volts at the terminals, and a current of 1·7 ampères, or about 6 kgms. per second (= 2·24 foot-pounds per second).

The next size larger, about the size of a small filbert, gives 2 to 3 candles with 4·2 volts at the terminals, and a current of 1·5 ampères. We do not know the life of these lamps under these conditions, which must be considered as excessive, for as the candle power in respect to energy

S

absorbed is as least as high in these little lamps as in the larger patterns, it is evident that they must be rather pushed, and must therefore last for a much shorter time.

ELECTRIC TELEGRAPHY.

Telegraph systems always include at least four elements: (1) the generator of electricity; (2) the transmitter; (3) the line; and (4) the receiver.

We need say nothing here about the generator of electricity, which is almost always a battery. The only batteries which are now used for telegraphs are the *Daniell* battery and the different forms of *gravity Daniell*, the *Leclanché*, and the bichromate battery (*Fuller*, *Higgins*, etc.). The *transmitter* and *receiver* are considered together under the name of *transmitting instruments*. The line, or, rather, the different lines we are about to examine are subdivided, according to their nature, into overhead, underground, and submarine lines.

OVERHEAD LINES.

Conductors.—Galvanised iron wire is used everywhere. In Germany it is sometimes covered with linseed oil. In England, where the line is exposed to corrosive vapours, the galvanised wire is immersed in a mixture of tar and bitumen, and is covered with a double layer of tarred line. In America a compound wire is also used, formed of a tinned steel wire, covered spirally with a ribbon of copper, and passed through a bath of tin to solder the copper to the steel. Compound wire is tenacious, light, and resists the action of vapours; but is expensive, and the copper coating sometimes comes off. Since 1877, a wire formed of a steel heart, electrically covered with copper, has been used in America.

Joints.—In France a sleeve is employed, and the Britannia joint in other countries. The solder run into the iron sleeve is composed of two parts of tin and one part of lead.

Insulators.—The double bell insulator is most often used, made of pure kaolin or porcelain clay; stone-ware and earthenware are also much used. The insulator is entirely glazed, except the edge on which it rests during the baking, which is carefully polished. Sometimes glass or ebonite insulators are used.

Insulation of overhead lines (*Culley*).—Insulation of a line 448 kilomètres long; iron wire 6 millimètres in diameter; resistance of conductor, 2,260 ohms.

Insulation per kilomètre in comparatively fine weather . . . . 21,920,000 ohms.
Insulation per kilomètre in wet weather, signals still good . . . . . . 307,000 ,,
Insulation per kilomètre, weak signals . . 291,000 ,,

The insulation should never be less than 300,000 ohms per kilomètre.

Calculation of the insulation resistance of an overhead line (*Varley*).—When the resistance is the insulation resistance of a line supported on equidistant posts, the current received C_r may be calculated in terms of the current sent C_s by the following formula:

Let

m be the resistance of the line between two posts;
i the insulation resistance at each post;
n the number of posts;
e the number $e = 2\cdot 718$.

$$\text{Let } Z = e^{n\sqrt{\frac{m}{i}}}.$$

Calculate the value of Z, then

$$C_r = \frac{2C_s}{Z + \dfrac{1}{Z}}$$

In the case of a well-insulated line, the ratio $\frac{m}{i}$ should be less than $\frac{1}{80{,}000}$.

This formula is only applicable to lines on which the highest potential used does not exceed 100 volts.

Leakage losses.—In practice, *Hughes* (1864) considers that on a line 300 miles long, with mean insulation, the current received is equal to one-third of the current sent out. For a line of the same length, *Prescott* considers that the received current varies between ·75 and ·2 of the current sent out, according to the insulation of the line.

Received current testing.—*Post-Office technical instructions.*—At both the sending and receiving ends of the line, resistance coils RR, of 10,000 ohms each, are to be included in the circuit when the received currents are measured. The battery sending the current is in all cases (whether on a short or long section) to be 50 Daniell's cells, and a supplementary battery of 10 cells with extra terminals (one connected to

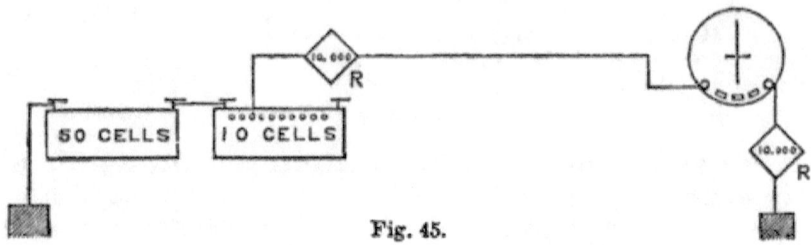

Fig. 45.

each cell) is to be included with the 50 cells as in Fig. 47, so that the electromotive force of the current can always be made approximately equal to 50 good cells by adding on one or more extra cells when the electromotive force of the battery diminishes.

The tangent galvanometer must be adjusted by means of the adjusting

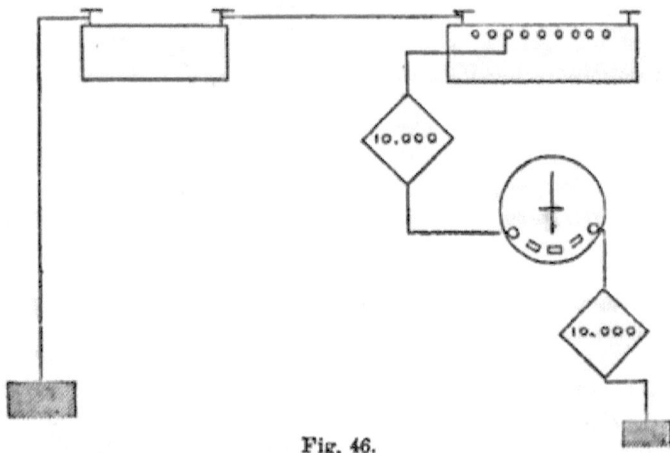

Fig. 46.

magnet, so that the standard cell (in good condition) with *both* plugs of the galvanometer *out*, gives a deflection of 25°.

The strength of the received currents (which must be measured with the *left*-hand plug *in*) is then obtained from the table on page 278. This table gives also the absolute insulation resistance in ohms of the line, corresponding to the strength of the received current; thus, if the received current gives 30°, then the strength of the latter will be 1·24 milliampères, and the total insulation resistance of the line 5,680 ohms, which, multiplied by the length of the circuit in miles, gives the insulation resistance per mile.

In order to test whether the sending battery is of the current power, the latter should be joined, as in Fig. 48, direct through the tangent galvanometer at the sending station with the two 10,000 ohms resistance blocks on circuit, and *both* plugs of the galvanometer out. If the battery is in proper order a deflection of 49½° should be obtained; if this is not the case the necessary number of cells from the 10-cell supplementary battery should be added, so as to obtain the required deflection.

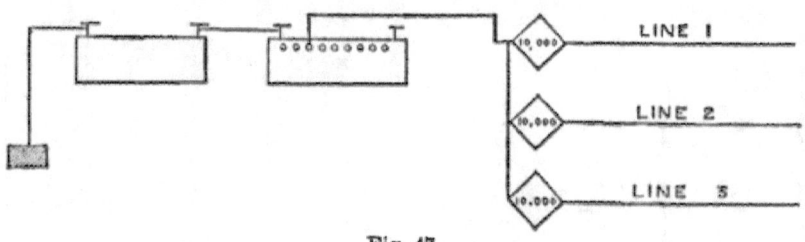

Fig. 47.

In cases where three sets of wires are being tested at the same time, *one* sending battery may be used for sending the currents on the "Universal system," the 10,000 ohms resistance blocks being interposed in each lead wire as shown in Fig. 49.

Office wire.—Annealed copper, 1 millimètre in diameter; two layers of guttapercha, separated by a layer of Chatterton's compound. Total diameter, 3 millimètres; insulation resistance, 50 megohms per kilomètre.

Earth and earth wires.—Every instrument (receiver, lightning guard, or battery) is provided with an earth wire, consisting of a copper wire, 1·6 millimètres in diameter. All these wires, brought together into a cable, form an earth cable, which thus has no sensible resistance. The earth cable ought to be soldered to as large a conducting-surface as possible, embedded in soil which is perfectly moist at all seasons of the year, and acts as a conductor over a large space, a pump, or an iron water or gas-pipe. The earth cable is soldered to both systems of pipes, if possible. The conducting wire ought to be soldered to the earth-plate, and the point of the junction carefully painted, so as

TABLE SHOWING STRENGTH OF RECEIVED CURRENTS AND EQUIVALENT INSULATION RESISTANCES.

Deflection.	Strength of Received Current.	Equivalent Insulation Resistance.	Deflection.	Strength of Received Current.	Equivalent Insulation Resistance.
Degrees.	Milliampères.	Ohms.	Degrees	Milliampères.	Ohms.
49	2·47	—	29	1·19	5,010
48½	2·42	271,000	28½	1·16	4,820
48	2·38	142,000	28	1·14	4,630
47½	2·34	96,500	27½	1·12	4,450
47	2·30	72,100	27	1·09	4,290
46½	2·26	57,500	26½	1·07	4,120
46	2·22	47,800	26	1·05	3,970
45½	2·18	40,700	25½	1·02	3,820
45	2·14	35,400	25	1·00	3,680
44½	2·11	31,200	24½	·977	3,540
44	2·07	27,900	24	·954	3,400
43½	2·04	25,200	23½	·932	3,280
43	2·00	22,900	23	·910	3,150
42½	1·97	21,000	22½	·888	3,030
42	1·93	19,300	22	·866	2,920
41½	1·90	17,900	21½	·845	2,810
41	1·87	16,600	21	·823	2,700
40½	1·83	15,500	20½	·802	2,600
40	1·80	14,500	20	·780	2,500
39½	1·77	13,600	19½	·759	2,400
39	1·74	12,800	19	·738	2,300
38½	1·71	12,000	18½	·718	2,210
38	1·68	11,400	18	·697	2,120
37½	1·65	10,800	17½	·676	2,040
37	1·62	10,200	17	·655	1,950
36½	1·59	9,700	16½	·635	1,870
36	1·56	9,220	16	·615	1,790
35½	1·53	8,780	15½	·595	1,710
35	1·50	8,370	15	·574	1,638
34½	1·47	7,990	14½	·555	1,560
34	1·45	7,630	14	·537	1,490
33½	1·42	7,290	13½	·516	1,420
33	1·39	6,980	13	·495	1,360
32½	1·37	6,680	12½	·475	1,290
32	1·34	6,400	12	·456	1,230
31½	1·31	6,140	11½	·436	1,160
31	1·29	5,890	11	·417	1,100
30½	1·26	5,650	10½	·397	1,040
30	1·24	5,430	10	·378	977
29½	1·21	5,210	9½	·359	919

Deflection from standard cell through galvanometer with both plugs out to be made 25°.

The battery sending current to give 49½° on galvanometer with both plugs out, and adjusted as above, with 20,000 ohms in circuit.

to prevent the joint from deteriorating. Leaden gas-pipes ought never to be used as earths. The cables of earth wires ought to be kept at least 2 to 3 centimètres from lead pipes, because, during a thunder-storm, a discharge might take place between the earth wire and the pipe. Instances are on record of the lead being melted, and the gas catching fire. The earth wire, however, may be connected to leaden gas-pipes outside the office, and in an open space where it is easy to see from time to time whether the soldering keeps in good order. If pipes are not available, a plate of galvanised iron, one mètre square, may be used, buried in damp earth, or placed in running water or a spring well (not a cistern).

The plate ought to be buried flat, and not rolled in a cylinder or spiral; it ought to be placed upright rather than horizontal. The earth in which the plate is buried ought to be moist at all seasons of the year. In hot countries, the earth where the plates are buried ought to be frequently watered.

For distances of less than one kilomètre, it is better to employ a return wire than earth plates. This return wire need not be insulated.

The resistance of a good earth ought never to exceed 10 ohms. Instead of a plate, a cable may be used, formed of iron wires 3 millimètres in diameter, which are frayed out and made to radiate in all directions under the earth. The covering of a submerged cable makes an excellent earth. The rails of a railway may also be used, on account of their great surface.

UNDERGROUND LINES.

Germany.—Cable of four to seven insulated copper conductors, coated with guttapercha, a layer of tarred Russian hemp, a covering of galvanised iron wires as a mechanical protection, and, lastly, a coating of asphalte as a protection against moisture.

France.—The underground cables are of two patterns, corresponding to iron wires of 5 and 4 millimètres.

The conductor is a twist of seven copper wires covered with two layers of guttapercha, with Chatterton's compound between them. The core thus formed is served over with tarred cotton, then several similar cores are twisted into a cable and covered with three envelopes: 1st, cotton ribbon; 2nd, a serving of line; 3rd, ribbon. Those conductors which are to be placed underground are simply covered with these envelopes and tarred. Those which are to be put up in tunnels or sewers are not tarred, but are introduced into leaden tubes 1·25 millimètres thick. The tubes are at least 50 mètres long without join. The serving and cotton ribbon are always steeped in a solution of sulphate of copper, even those which are afterwards to be tarred. The use of gas-tar is prohibited. The copper used has a conductivity of 80 per cent. of that of pure copper. The insulation must not be less than 400 megohms at 20° C. The cables are manufactured in lengths of 400 mètres, with a maximum allowance of at most 1·5 mètres. All the conductors of any one section are without join.

UNDERGROUND CABLES WITH SEVEN CONDUCTORS.

Diameter of Strands of the Copper Twist.	Diameter of Conductor covered with Guttapercha.	Diameter of the Cable.		Price of Cable per Mètre.	
		Braided.	Leaded.	Braided.	Leaded.
mm.	mm.	mm.	mm.	fr.	fr.
·7	5·1	20	22·5	1·80	2·45
·5	4·5	18	20·5	2·65	2

Various forms of cable.—Cables enclosed in lead are better protected than those coated only with tarred fibre; but by diminishing the volume of the tube, so as to avoid inconvenience in handling, and expense, the inductive capacity of the cable is increased. Attempts have been made to employ a dielectric of less capacity, such as guttapercha, indiarubber, resin, paraffin, petroleum, nigrite, ozokerit, etc.

Berthoud and Borel's cable.—A copper conductor covered with one or two layers of cotton passed through a mixture of resin and paraffin. Several wires are united and inclosed in a leaden casing. The tube, after being passed through thick turpentine, receives a second leaden tube. The insulation is as much as 30,000 megohms per kilomètre at the ordinary temperature.

Brooks' cable.—The conductors are covered and separated by a layer of perfectly dry jute or hemp. The conductors are introduced into iron tubes jointed together by screwed joints and filled with petroleum oil. A tube 4 centimètres in diameter will hold as many as 50 telegraph wires. The insulation is less than that of ordinary cables. The capacity is ·2 microfarads per kilomètre.

SUBMARINE LINES.

Submarine conductors are called cables. A cable is always composed of three parts: 1st, the conductor, always copper; 2nd, the dielectric or insulator; and, 3rd, the protecting envelope or armature. The conductor and the insulator together form the core of the cable. The conductivity of the copper varies between 94 and 98 per cent. The resistance of the conductor, according to its length, from 4 to 12 ohms per knot; the insulation resistance from 250 to 700 megohms per knot; and the electrostatic capacity from ·35 to ·28 microfarad per knot. The points to be aimed at in the construction of a cable are to obtain the smallest possible resistance of the conductors, the smallest possible electrostatic capacity, and the greatest possible insulation resistance. Let D be the diameter of the core, d the diameter of the conductor; the maximum speed of transmission is when

$$D = d\sqrt{e} = 1.649\, d.$$

Mechanical considerations make it necessary to increase the ratio $\frac{D}{d}$ which varies from 2·4 to 3·4 in practice. In the table on page 282 will be found the details of the principal cables of recent construction. We must refer the reader for more full information on this very special question to the works indicated in the bibliography placed at the end of this volume.

INSTRUMENTS.

In a telegraphic system, what are called the instruments are the transmitters, receivers, galvanometers, lightning protectors, relays, etc., which are used in the different operations.

Classification.—The nature of the signals received enables us to divide the instruments into several groups:

1st. *Visual signal instruments.*—Transmission is effected by a series of signals which leave no trace. Single needle instruments (still much used in England): Wheatstone's A B C instrument, Breguet's telegraph, and Sir W. Thomson's reflecting galvanometer.

2nd. *Acoustic instruments.*—*Sounders*, very much used in America, and in the English military service; the despatch is read by the Morse alphabet.

3rd. *Registering instruments.*—The despatch is marked on a continuous band of paper in conventional characters; Morse instrument, siphon recorder.

4th. *Printing instruments.*—The despatch is printed in ordinary characters on a continuous band of paper; the type is the Hughes' telegraph.

5th. *Automatic instruments.*—Reproducing at a distance writing and drawings (Caselli, Meyer, Lenoir, Edison) are not used in practice.

6th. *Articulating instruments or telephones.*—(*See* page 289.)

DETAILS OF SOME RECENTLY LAID SUBMARINE CABLES.

(The Telegraph Construction and Maintenance Co.)

Cable and Date of Laying.	Length in Knots.	Electrical Details at 24° C.				Weight per Knot in Pounds and Tons.						Nature of Cable.
		Conductor Resistance per knot in ohms.	Conductivity pure copper = 100.	Dielectric Resistance per knot in megohms.	Electrostatic capacity per knot in microfarads.	Copper lbs.	Guttapercha lbs.	Iron tons.	Hemp tons.	Asphalte tons.	Total tons.	
Placentia to Saint-Pierre (1872).	110	11·93	93·4	378	·303	107	140	2·344	·153	—	2·607	Deep sea.
						107	140	9·845	·519	1·145	11·619	Shore end.
Alexandria to Crete (1873).	359	11·78	94·6	335	·307	107	140	9·845	·519	1·145	11·619	Shore end.
						107	140	5·917	·325	·691	7·043	Shoal water.
						107	140	1·038	·075	·337	1·56	Deep sea.
Ireland to Newfoundland (1874).	1837	3·135	95·1	282	·332	400	400	17·46	1·684	—	19·5	Shore end.
						400	400	3·55	·164	·408	4·122	Shoal water.
						400	400	·662	·147	1·08	2·246	Deep sea.
Yankalia to Kingscote (1875).	38	11·641	95·8	228	·3	107	140	9·845	·650	1·120	11·725	Shore end.
						107	140	5·325	·4	1·051	6·866	Shoal water.
						107	140	2·7	·095	·635	3·54	Deep sea.
Ireland to Newfoundland (1880).	1423	4·161	95·5	478	·315	300	300	·86	·115	·725	1·967	Deep sea.
						300	300	·72	·115	·705	1·807	Deep sea.
Valentia to Greitseil (1882).	841	9·098	100	567	·354	130	130	11·9	·55	·4	12·935	Shore end.
						130	130	5·6	·35	·633	6·729	Shoal water.
						130	130	2·78	·125	·499	3·52	Deep sea.

High-speed instruments enable the whole transmitting power of the line to be utilised, which is always much greater than the speed of transmission of the most skilful clerk. They work by multiplying the number of operators working on the same line. In the automatic instruments a certain number of operators perforate bands of paper which afterwards pass through the transmitter, and produce very rapid emissions of current, which are registered at the receiving end on a continuous band of paper like that of the Morse instruments.

In duplex instruments two operators transmit the despatches simultaneously in opposite directions on one and the same wire. In the diplex the two operators transmit simultaneously in the same direction. In the quadruplex two operators at each end of the line transmit simultaneously four despatches; two in one direction, and two in the other.

Lastly, in the multiplex instruments based on the division of time by means of synchronism established between the sending and receiving stations, the line is placed for equal and equally divided fractions of time successively in communication with several groups of operators, who take advantage of the interval of time between two successive communications to prepare the next signal. These numerous high-speed instruments are not only very different in their principle, but may also be distinguished from one another by the nature of the signals they transmit. Thus the duplex, diplex, and quadruplex instruments generally work sounders. The Wheatstone transmits Morse signals (exclusively), which are recorded; whilst the synchronous instruments transmit sometimes Morse signals, as in *Meyer's* instruments; sometimes ordinary characters, as in *Baudot's* instrument, which is a marvel of mechanical ingenuity. Space preventing us from giving a description of all these instruments, we will only point out their general arrangements, and the working of their principal parts.

Electro-magnets.—*Specification of French telegraph service.*

Soft iron cores and armature perfectly annealed, must retain no appreciable traces of magnetisation after a prolonged transmission of Morse signals with a battery of 100 Callaud elements; must not be touched with the tool after being annealed.

Wire of the coils.—Copper of conductivity above 90 per cent. covered with white or cream-coloured silk. The fine wire to be in one piece, one of the ends soldered to the reel if it be of brass; outside layer No. 16 wire of ·44 millimètre. The finished coil to be covered with asphalte varnish, allowance 2 per cent. The difference between two coils must be less than 4 ohms, but the number of turns must be the same.

Hughes' instrument.—Minimum carrying power of the magnet 5 kilos, coils of the electro-magnet of number 32 (·17 millimètre bare). minimum number of turns 11,000, maximum total thickness 11 millimètres, maximum resistance 600 ohms. Hollow soft iron core 1·5 millimètres thick, 6 millimètres external diameter, 6 centimètres long, and filled up to 3 millimètres of thickness to receive the screw fixing the pole piece. Testing speed ninety to 140 revolutions of the chariot per minute without slowing.

Morse receivers.—Clock-work movement to go for forty minutes, speed of unrolling ribbon 1·7 mètres to 1·4 mètres per minute, seven thousand turns of No. 29 wire (·21 millimètre bare); resistance less than 250 ohms, thickness of the coils not to exceed one centimètre.

Relay speakers.—Coil mounted on a brass reel; in the centre, 2,000 turns of No. 36 wire ·12 millimètre bare, then 5,000 turns of No. 34 wire ·14 millimètre bare, in all 7,000 turns, minimum resistance 500 ohms.

Reverse current coils.—Each coil 4,500 turns of No. 32 wire (1·7 millimètres bare), resistance 2,500 ohms. Bells with lightning guard, each coil 2,000 turns of No. 32 wire; resistance 50 ohms, allowance 5 ohms. Maximum distance between the edges of a lightning guard half a millimètre.

Methods of diminishing the extra current on breaking circuit.—The use of these methods is especially necessary when powerful currents are used. (*a*) Arrange a coil of as fine wire as possible as a shunt, having a resistance forty times that of the electro-magnet. The wire to be coiled half from left to right, and half from right to left, so as to prevent the formation of extra currents in the shunt (*Dujardin*). Place a small condenser of tinfoil between the terminals of a receiver; the capacity of this condenser must vary with the resistance of the electro-magnet and the power of the battery, generally from one-eighth to a quarter of a microfarad (*Culley*). When it is possible to arrange two coils of an electro-magnet for quantity or in parallel circuit, the extra currents which are produced in each of the coils tend to neutralise each other. This arrangement applied to the Wheatstone instrument has enabled the rapidity of transmission to be increased from 10 to 20 per cent. (*Preece*).

Strength of telegraphic currents.—Distinction must be made between the working current and the mean current. The working current is that sent out at the sending end, and measured at that end, which in France varies between 12 and 20 milliampères. The received current which works the receiver is diminished by the influence of leakage, and according to the state of the line and the atmosphere, the length of the line and its insulation, etc., varies between ·7 and ·2 of the initial current. The *mean current*, which varies between 7 and 13 milliampères, represents the continuous current, which during the twenty-four hours would produce the same expenditure of chemical action as the daily work of the working current. The quantity of electricity expended per twenty-four hours varies between 500 and 1,200 coulombs, and represents a daily deposit of from ·2 to ·4 gramme of copper. The current which actually passes through the receiver varies between 3 and 8 milliampères.

Mean strength of telegraphic currents used in India (*Schwendler*).

Working currents during the dry season, eight months, in milliampères	6·4
During the rainy season, four months, in milliampères	13
Local circuit working a speaker, in milliampères	·72
Resistance of the speaker, in ohms	25 to 35
Siemens relay.—Resistance in ohms	500
Strength necessary to work the instrument, in milliampères	2

Range of electro-magnet receivers (*Schwendler*).—

Let c be the weakest which will work a telegraph instrument, C the strongest current which it will bear, its range is:

$$Range = \frac{C}{c}.$$

It is always as well that the range should be as high as possible.

The range of an electro-magnet receiver is a decreasing function of the speed. The following is a table of the experiments made by Schwendler on a Siemens polarised relay to determine its variations:

Number of Experiment	Number of Contacts per Minute.	Weakest Current in Milliampères C.	Strongest Current in Milliampères C.	Range $\frac{C}{c}$
1	53	·89	14·35	16·1
2	101	1·03	14·35	14
3	138	1·14	14·35	12·6
4	313	1·81	14·35	7·9
5	419	1·81	8·36	4·6

Four hundred and thirty contacts per minute correspond to a speed of 20 words of 5 letters per minute, the word Paris being taken as a type. The range of a relay working at this speed cannot **exceed 4**. The relay experimented on above will work with currents varying between 2 and 8 milliampères without changing its adjustment.

Sensitiveness of a telegraphic instrument.—The wire ought to be so arranged as to produce the most intense possible **magnetic** field at that part where the work is to be done. The movable parts **ought to** be as small as possible, so **as** to diminish their moment of inertia, and because a small magnetic field costs less to produce than a large field of equal intensity. Those parts which are magnetised by the current ought to magnetise and demagnetise rapidly; that is to say, have but little magnetic inertia, have a small mass and as little coercive force as possible. All contact between the armature and the electro-magnet should be avoided because of residuary magnetism.

Siemens relays (*Schwendler*).—The resistance R of a relay may be deduced from the formula:

$$R = \frac{5}{8} L.$$

L being the resistance of the longest line on which the relay is required to work.

Range.—A Siemens relay ought always to have a higher range than 25; many have a range of 35, and some instruments have as much as 55. A Siemens relay of 500 ohms resistance works well with a current C of 2 milliampères. The same **relay** having a resistance of r' would work with a current c':

$$c' = \frac{c\sqrt{r}}{\sqrt{r'}} = \frac{44\cdot 8}{\sqrt{r'}} \text{ milliampères.}$$

This approximate formula does not take the **thickness of** the insulator into account.

Local sounders.—In India they have a mean of from 25 to 35 ohms resistance, and work with four Minotti elements in series, and a current of 72 milliampères.

Portable sounders.—Polarised relay 500 ohms resistance, 250 for each coil.

TELEGRAPHIC INSTRUMENTS. 287

Dial telegraph.—Each turn of the handle represents 13 emissions of current and 13 interruptions. A well-constructed instrument gives two turns and a half of the needle per second; in practice one turn per second is counted. Each letter requires as a mean one-half turn and a wait of half a second. The speed is 60 letters per minute, or 10 words, each word comprising 5 letters, and one return to the $+$, which indicates the end of the word.

Morse instrument.—*Spacing and length of the signs.*—One dash is equal to three dots; the space between the signs of the same letter equal to one dot; the space between two letters equal to three dots; the space between two words equal to five dots; mean word, five letters.

Speed of transmission.—A good clerk, 18 to 20 words per minute; mean, 12 to 18.

Speed of the receiver.—Depends upon the number of dots per minute which the armature can make, and varies from 800 to 2,000. The letter of mean length is r (- — -). The length of a dot which can be easily read is three-quarters of a millimètre. If the strip unrolls at the rate of 1·2 mètres per minute, as many as 32 words per minute may be transmitted; beyond that speed, the rate of unwinding of the slip must be increased.

Mean work.—25 simple despatches of 20 words plus the address (say 30 words) per hour, or 750 words or 3,750 letters per hour, the maximum of a letter being 4 dashes. The number of dashes produced is almost 15,000 to the hour, or 5 per second.

MORSE ALPHABET.

a	· —	j	· — — —	s	· · ·
ä, æ	· — · —	k	— · —	t	—
b	— · · ·	l	· — · ·	u	· · —
c	— · — ·	m	— —	ü, ue	· · — —
d	— · ·	n	— ·	v	· · · —
e	·	ñ	— — · — —	w	· — —
é	· · — · ·	o	— — —	x	— · · —
f	· · — ·	ö, œ	— — — ·	y	— · — —
g	— — ·	p	· — — ·	z	— — · ·
h	· · · ·	q	— — · —	ch	— — — —
i	· ·	r	· — ·		

APPLICATIONS OF ELECTRICITY.

Full stop		— — —
Colon	(:)	— — · · — —
Semicolon	(;)	— — · — · —
Comma	(,)	· — · — · —
Note of interrogation	(?)	· · — — · ·
Note of admiration	(!)	— — · · — —
Hyphen	(-)	— · · · · —
Apostrophe	(')	· — — — — ·
Parenthesis	( )	— · — — · —
Inverted commas	(" ")	· — · · — ·

NUMERALS.

1	· — — — —	6	— · · · ·
2	· · — — —	7	— — · · ·
3	· · · — —	8	— — — · ·
4	· · · · —	9	— — — — ·
5	· · · · ·	0	— — — — —

Bar of division	— — — — —
Call signal	— · — · — ·
Understand message	· · — ·
Repeat message	· — · · — · ·
Correction, or rub out	· · — — — — ·
End of message	· — · — · —
Wait	· — · · ·
Cleared out, and all right	· — · · — · · —
Begin another line	· — · — · ·

The Hughes' instrument.—The printing axle turns seven times quicker than the chariot and the type axle. The keys, which are pressed down successively, must be separated by an interval of four keys at least, as the number of keys is 28. The velocity of the type-wheel varies from 40 to 150 revolutions per minute; the mean is 110 to 124 revolutions. 1·54 letters are transmitted per revolution, or 185 letters per minute if the chariot perform 120 revolutions. Two letters per revolution of the chariot are generally counted, each word being composed of five letters and a blank. Thirty-one words per minute is the rate at 120 revolutions.

The contact piece on the chariot covering three divisions of the contact box, the contacts last for ·053 second. The mean rate of working is 45 to 50 despatches per hour; 60 are often sent on short lines.

Wheatstone's automatic transmitter.—One operator can perforate 25 despatches in an hour. On a short line the instrument

TELEPHONY.   289

can transmit 130 words per minute. Between Paris and Marseilles, 863 kilomètres, the mean rate of sending is 85 despatches to the hour with five clerks at each end. A series of ten despatches passes through the transmitters in five minutes. The emissions of current vary between 10 and 90 per second.

Speed of transmission of telegraphic instruments.—These figures show the rates for despatches of a mean of 20 words on a line of 600 to 700 kilomètres in length per hour :

Morse 25	Wheatstone 90
,, duplexed 45	,, duplexed . . . 16
Hughes 60	Reflecting galvanometer . . 36
,, duplexed 110	,, ,, duplexed 50
Meyer per key-board . . 25	Gray (reading by sound) per office 3
,, with four key-boards . 100	Syphon recorder 25
Baudot per key-board . . 40	,, ,, duplexed . 35
,, with four key-boards . 160	Foot, 1000 words per minute between
,, with six key-boards . 240	Boston and New York (460 kilomètres).

TELEPHONY.

Telephony is the transmission of articulate sounds to a distance. A telephonic system always includes :

(1) A *transmitter*, which converts articulate sounds characterised by undulatory vibrations into undulatory currents ;

(2) The *receiver*, which transforms the undulatory currents from the transmitter into undulatory vibrations, similar to, but not identical with, the vibrations which have affected the transmitter ;

(3) A *line* formed of one wire or two wires joining the instruments.

Transmitters are divided into two classes :

1. *Magnetic transmitters*, requiring no battery. The sonorous waves produce undulatory currents, which afterwards act on the receiver.

2. *Battery transmitters*, carbon transmitters, microphones, etc. The sonorous waves modify the current furnished by a constant independent source (battery or accumulator).

MAGNETIC TRANSMITTERS.—All magnetic transmitters are reversible. They produce undulatory currents under the action of articulate sounds, and reciprocally reproduce the corresponding articulate sounds from undulatory currents passing through them. The type is the *Bell* telephone, which is formed of a plate of thin sheet-iron vibrating in front of a magnet, round which a coil of insulated conducting wire is wound. The

T

vibrations of the plate produce induction currents in the coil. The strength and direction of the currents are directly connected to the movements of the vibrating plate. Many modifications in its form have been introduced, which in no way affect the principle discovered by Prof. Bell.

BATTERY TRANSMITTERS.—Based on the variations of resistance of carbon under the influence of pressure, which was discovered by Du Moncel in 1856, and applied to telephony for the first time by Edison in 1877. The resistance of an imperfect contact was used for the first time by Hughes, in 1878, and the inventor named the apparatus the microphone. All transmitters now used are based on the variation of resistance of imperfect contacts, the contact being made to vary under the influence of articulate sounds. Carbon is the best substance to use, because it does not oxydise and is infusible, and also because it is of low conductivity, and its resistance decreases when heated.

The different forms of microphones differ by the number of imperfect contacts, their arrangement, their grouping, the nature of the sounds to be transmitted, etc. It is impossible to give any definite rules, and, as Mr. Preece has justly remarked, "the microphone at present defies mathematical analysis."

1. *In direct circuit*, for short distances.
2. *With induction coils.*—The undulatory current passes through the primary coil of an induction coil; the secondary circuit is joined to the receiver, which is thus affected by the induced currents. This last method alone is used for long lines and telephone systems. The resistance of the transmitter varies from 1 to 150 ohms. The resistance of the induction coils is also very variable, and no rule can be given, because of the secondary phenomena of self-induction, charging of the lines, etc.

Ader's microphone, used by the Société Generale des Téléphones in Paris, has a mean resistance of 5 ohms, the primary circuit of the coil 1·5 ohms, the secondary coil 150 ohms, and the receiver 75 ohms: the flexible conductor about 4 to 5 ohms. The mean strength of the inducing current does not exceed a quarter of an ampère, or the twentieth of an ampère per contact. In *Moser's* experiments, with 24 transmitters in parallel arc, the current was 24 ampères, or one ampère per microphone, one-fifth of an ampère per contact. The plate of the receiver is ·3 millimètre thick, the wire of the coils ·09 millimètre in diameter. The induction coil is made of wire, ·5 millimètre for the primary and ·14 millimètre for the secondary.

For the telephonic system of Paris, the primary circuit is formed of the transmitter, the coil, and three Leclanché cells, with agglomerate plates, each arranged in series.

Receivers.—The most used, in fact, the only ones used in practice, are:

Magnetic receivers.—The most simple and the best is the Bell telephone in its innumerable variations. Some are made with two poles, like the *Siemens* and *Gower*. The super-saturated telephone of *Ader* and the concentric pole telephone of *d'Arsonval*.

Various receivers.—Many other actions besides magnetic actions have been used in telephonic receivers. We will only cite the *electromotograph* of Edison, based on the variations of friction between chalk and platinum, moistened by a saline solution, under the action of a variable current; the mercury telephone of *Antoine Breguet*, based on electro-capillary actions; the heat telephone of *Preece*, based on the heating and expansion of a wire traversed by an undulatory current; *Dolbear's* electrostatic telephone, based on the reciprocal actions of two plates charged with variable quantities of electricity, etc.

Lines.—With very few exceptions, an earth return should never be used for telephonic transmission. A double wire or metallic circuit should be used. An earth return may, however, be used where there is no fear of induction.

Induction.—The noise produced by the action of neighbouring circuits on a telephonic circuit is thus called. It often prevents direct conversation from being heard. When the telephonic circuit runs near many telegraph wires, it often sounds exactly like the boiling of a kettle. Induction may be diminished to a certain extent by the following devices:

(1) By diminishing the sensibility of the receiver and increasing the transmitting currents, so as to weaken the external disturbances.

(2) By establishing an induction screen between the telephone wire and the other wires, by using an insulated wire covered with a metallic coating connected to earth.

(3) By modifying the causes of disturbance, by sending graduated currents instead of abrupt currents in the neighbouring circuits.

(4) Neutralising the effects by means of counter-induction instruments.

(5) Always employing a double or metallic circuit, placing the two wires close together; or, better still, using insulated wire, and twisting them together. This last method is much the most efficacious and the most employed.

Losses on the line.—The use of a double wire demands perfect insulation on the line; otherwise all exterior currents are felt in the instrument, especially when these external currents have an earth return. Telephonic lines with an earth return are also sensible to external disturbances, electric lighting, telegraphs, earth currents, storms, etc.

Distance.—When all disturbing causes are carefully guarded against, it is possible to telephone to a distance of 700 kilomètres, in perfect silence and with an overhead line.

Through submarine cables it has never been possible to exceed 180 kilomètres, because of the electrostatic capacity of the cable. No more than 40 kilomètres is possible with underground lines (January, 1883).

Work done by batteries used with microphones (*E. Reynier*).—These figures have been calculated from the observed expenditure of zinc in the Leclanché batteries in the busiest offices on the Paris telephone system (each office having a battery of 3 elements, one Ader microphone, and one induction coil).

Mean strength of inducing current	·084 ampère.
Work given out by the battery when the microphone is in action	·025 kilogrammètre per second.
Annual work of a very busy office, the instrument being spoken through 7 hours out of the 24	235,425 kgms.

From Reynier's calculations, the 3000 offices of the Paris system, each being supposed to be as busy as the central offices, would use daily 1,935,000 kilogrammètres, or *one horse-power for about 7 hours and 5 minutes*. With accumulators of efficiency 80 per cent., charged by dynamos of efficiency 80 per cent., *a driving power of one horse-power working 12 hours per day* would be ample for the work of the whole Paris system.

Simultaneous telephony and telegraphy on the same wire. *Van Rysselberghe's system* (1883).—After he had succeeded in stopping the induction produced by telegraph wires on telephone wires by graduating telegraphic currents by the use of resistances, condensers, and electro-magnets, Van Rysselberghe was enabled to connect telephones directly to wires used for telegraphic communication, and use the wires for the simultaneous transmission of Morse signals and articulate speech.

Fifth Part.

RECIPES, PROCESSES, ETC.

Fusible alloys (*Agenda du chimiste*).
D'Arcet's alloy, fusing at 94° C.

Lead	5 parts.
Tin	3 ,,
Bismuth	8 ,,

Wood's alloy, fusing between 66° and 71° C.

Lead	2 parts.
Tin	4 ,,
Bismuth	7 to 8 ,,
Cadmium	1 to 2 ,,

Fusible amalgam, fusing at 53° C.

D'Arcet's alloy	9 parts.
Mercury	1 ,,

Alloys used for instruments:

	Copper.	Zinc.	Tin.
Tombac, or white copper	86 to 88	14 to 12	—
,, yellow ,,	88·88	5·56	5·56
,, red ,,	91·66	8·34	—
Romilly's brass	70	30	—

Aluminium bronze:

Copper	90 parts.
Aluminium	10 ,,

Silvering for curved mirrors:

Tin	4 parts.
Mercury	1 ,,

Alloy for nickel coins (Germany, Belgium, United States)

Copper	75 parts.
Nickel	25 ,,

SOLDERS.

	Copper.	Zinc.	Tin.	Lead.
Hard solder, yellow, high melting point	53·3	43·1	1·3	·3
Hard solder, whitish yellow, low melting point	44	49·9	3·3	1·2
Hard solder, yellowish white, very low melting point	57·4	28	14·6	—
Hard solder, whitish yellow, very strong	53·3	43·7	—	—
Brass solder	1·5	6	Brass 10	
Plumber's solder	—	—	33	66
Tinman's solder	—	—	50	50

Low temperature solder.—For use when the parts to be soldered will not stand a high temperature. Finely divided copper (obtained by precipitating a solution of copper sulphate with zinc) is mixed with concentrated sulphuric acid in a porcelain mortar. 30 to 36 parts of copper are taken, according to the degree of hardness desired, and 70 parts of mercury are stirred in. When the amalgam has completely formed, it is washed with hot water till all traces of acid are removed. It is then allowed to cool.

When this composition is to be used it is heated until it is of the consistency of wax, so that the surfaces to be joined may be readily smeared with it. When cold they adhere very strongly.

To give copper the appearance of platinum.—Scale clean and dip in the following bath until the desired appearance is produced.

Hydrochloric acid	1 litre.
Arsenious acid	250 grammes.
Copper acetate	45 ,,

Dry by rubbing with blacklead.

Platinised silver.—Used in Smee's batteries. Dissolve a little bichloride of platinum in acidulated water, and decompose the solution by a current, taking a plate of platinum as the anode, and the silver plate to be platinised as the cathode. A rough deposit is thus obtained which facilitates the disengagement of the bubbles of hydrogen.

Platinised carbon (*Walker*).—The carbon plates are first purified by soaking them for some days in sulphuric acid diluted with three to four times its volume of water; a tinned copper conductor is then fastened to it by tinned copper rivets. The carbon is then platinised by electrolysis, the carbon plate being used as the cathode, the anode being either a platinum or carbon plate. The solution used is thus prepared: sulphuric acid, diluted with ten times its volume of water, is taken, and crystals of chloride of platinum are added until the solution becomes of a beautiful straw yellow colour. After the current has passed for about twenty minutes the plate is finished; it may be tested by using it as a cathode in the electrolysis of water; it ought to allow the hydrogen to escape freely without sticking to it in the form of bubbles.

Platinised iron.—*Paterson* dips the plate to be platinised into an acid solution of platinum in aqua regia.

Amalgamation of iron.—The iron is steeped for some time in a solution of a mercury salt, or in mercury covered with very dilute sulphuric acid. *Boettger* heats the iron in a porcelain vessel with a mixture of 12 parts mercury, 1 zinc, 2 sulphate of iron, 12 water, and 1·5 hydrochloric acid. The simplest and quickest way to amalgamate iron is to clean it well with dilute acid, rinse in clean water, and then rub it with an amalgam of sodium or potassium (*Nature*).

Amalgamation of zinc.—Put a little mercury in a plate or shallow dish, fill up with weak dilute sulphuric acid and water, and rub the zinc well with a pad of old rag, dipping the zinc from time to time into the mercury; when the surface looks quite silvery, wash well with clean water, and stand the zinc edge downwards to drain, putting a dish under it to catch the excess of mercury, which will drain off. (This mercury contains much zinc.)

To purify mercury.—If one of the new low-pressure distilling apparatus be not at hand, put the mercury in a deep vessel, put plenty of dilute sulphuric acid over it, and place a piece of carbon (a bit of an electric light carbon answers very well) into the mercury, weight it or tie it down so that there is good contact with the mercury; this arrangement sets up local action, and dissolves out all metallic impurities; do not carry the action too far, as you may dissolve some of the mercury in the form of mercury sulphates.

Silver black (*A. Bailleux*).—1st. Take nitric acid at 40°, and dissolve silver (coins will do) in it until it is *saturated*. 2nd. Gently heat the object to be blackened, which ought not to be joined with tin solder. 3rd. Dip the object into the silver solution until it is cold, then replace it on the fire to dry. It is then black. Allow it to cool, then rub it with a softish brush and blacklead.

Gilt plumbago (*Tabauret*).—For giving a conducting surface to electrotype moulds, ·10 gr. of chloride of gold is dissolved in one litre of sulphuric æther, 500 to 600 gr. of plumbago (in fine powder) is thrown in, the whole is poured out into a large dish, and exposed to air and light. As the æther evaporates the plumbago is stirred and turned over with a glass spatula. The drying is finished by a moderate heat, and the plumbago put by for use.

Cyanide of potassium.—There are several qualities. No. 1 contains from 96 to 98 per cent. of pure cyanide. No. 2, for copper and brass platers, 65 to 70 per cent. No. 3, used by photographers, from 40 to 50 per cent.

Chloride of gold.—1 gramme of metallic gold corresponds to 1·8 grammes of neutral chloride, and to 2 or 2·2 grammes of the acid chloride such as is obtained from chemical manufacturers.

Porous pots.—Minimum leakage with distilled water at 14° C. 15 per cent. in twenty-four hours.

Morse paper.—Width: 1 centimètre. Weight: 53 grammes per 100 mètres; breaking strain: 1,300 grammes.

Sulphate of copper ought to contain less than 1 per cent. of iron, and 24 per cent. of pure copper ($CuSO_4 + 5H_2O$).

Ammonium chloride or **sal-ammoniac** (NH_4Cl) ought to contain less than 1 per cent. of impurities, and less than five thousandths of lead salts.

Dextrine dissolved in four times its weight of water ought to produce complete adhesion between two pieces of paper in ten minutes without sensibly discolouring them.

Black oxide of manganese.—Without dust of the kind called *needle* manganese; it ought to contain at least 85 per cent. of manganese peroxide.

Mean composition of commercial sulphate of copper (*Culley*):

Crystallised copper sulphate	99·66 to	98·48
Iron sulphate	·09 ,,	·12
Water	·35 ,,	1·4

Mean composition of certain samples of commercial zinc (*Culley*):

Zinc	99·27	98·76	97·85	98·89
Lead	·67	1·18	2·05	1·13
Iron	·06	·06	·1	·02

The purest sample was Silesian zinc. Rolled or drawn battery zincs ought to contain at least 98·5 per cent. of pure metal, and at most half per cent. of iron.

Purification of common commercial sulphuric acid (*d'Arsonval*).

—It may be purified by merely shaking it up with common lamp oil in the proportion of 4 or 5 cubic centimètres of oil to the litre of acid. The foreign bodies which would attack the zinc, arsenic, lead, etc., are precipitated.

Gilder's verdigris.

—Gilders call a bath of nitric acid in which they clean copper and its alloys, *strong water*. When the strong water is nearly saturated with copper it ceases to *bite*. It is refreshed by adding sulphuric acid, which forms an impure sulphate of copper improperly called verdigris, and sets the nitric acid free.

The composition of this verdigris is rather variable. The following are the results of an analysis made by M. *Van Heurck* of a sample from a Paris shop.

Water evaporable at 15° C. (moisture and water of crystallisation)	31·4
Substances volatile at a red heat (water of combination and a little nitric acid)	9·1
Oxide of copper	30·2
Sulphuric acid	29·3

Normal sulphate of copper contains:

Water	36·3
Oxide of copper	32·32
Sulphuric acid	31·38

It may be seen that the *verdigris* may be used advantageously instead of sulphate of copper in batteries of the Daniell class. The more so because the impurities increase the conductivity of the fluid. The verdigris costs about 45 per cent. less than sulphate of copper. More than one hundred thousand kilogrammes per year are produced in Paris (*E. Reynier*).

Purification of graphite (*Pelouze* and *Fremy*).—Graphite may be purified by reducing it to coarse powder, and mixing it with about one-fourteenth of its weight of chlorate of potash. The mixture is well mixed with twice as much concentrated sulphuric acid as graphite in an iron vessel, and heated in a sand bath until all the vapours of chlorous gas have ceased; when cold it is thrown into water and well washed. The washed and dried graphite is then heated red hot; it increases considerably in volume, and falls into a very fine powder. To purify it completely it must be levigated. After this operation it may be considered to be pure graphite suitable for a number of commercial purposes.

Magnetic figures may be obtained by placing a piece of slightly gummed paper over a magnet, and throwing iron filings on to it, and tapping the paper. Much better figures are obtained with sheets of glass covered with gum, and dried when the figure is formed; if the glass be exposed to steam the gum softens, and on drying fixes the filings in place; such glass plates may be used in the magic lantern for lecture purposes.

Soldering wires (*Culley*).—To solder iron wires together dissolve chloride of zinc (or kill spirits of salts with zinc), add a little hydrochloric acid (spirit of salt) to clean the wire. The rain soon washes off the excess of chloride of zinc. To solder iron and copper wires together the excess of chloride must be washed off, and the joint covered with paint or resin, or solder with resin.

For *unannealed* wires solder at as low a temperature as possible.

The zinc solution, or spirit of salt, should never be used except for overhead out-door lines. All joints in covered wire, whether run underground or above ground, and all joints within doors, either in covered or uncovered wire, should be made with resin. No spirit of salt, either pure or killed with zinc, should ever be allowed in an instrument maker's shop or dynamo factory. Workmen will use it, if not watched. Its presence may often be detected by holding an open bottle of strong solution of ammonia (liquor ammoniæ) under a newly made joint, if it becomes surrounded with a slight white cloud or mist, spirit of salt in some form has been used.

Joints of guttapercha-covered wire (*Culley*).—Exact perfect cleanliness. Remove the guttapercha for about 4 centimètres, clean the wire with emery paper, twist the wires together for about 2 centimètres, cut the ends off close so as to leave no point sticking out. Solder with resin and good solder containing plenty of tin. The guttapercha is then split, and turned back for about 5 centimètres, the soldered joint is covered with Chatterton's compound, and the guttapercha on each side of it is warmed and manipulated until the two sides join. The joint is finished with a hot soldering iron, taking care to smooth it off well, without burning it; it is then covered with another layer of Chatterton's compound. A sheet of guttapercha is then taken, warmed at a spirit lamp, and drawn out carefully so as slightly to diminish its thickness. Whilst both guttapercha and Chatterton's compound are warm the sheet is laid on the joint, and moulded round it with the thumb and forefinger. The joint is then trimmed with scissors; the edges kneaded in and smoothed down with a hot iron. When the joint is cold, another coating of Chatterton's compound is applied, and covered with a longer and broader piece of sheet guttapercha. The whole is then covered with a final coating of Chatterton's compound, spread with the iron, and polished by hand when cold, taking care to keep the hand well moistened. It is indispensable to obtain intimate and perfect union between the new guttapercha and that which covers the wire. A much neater and cleaner joint can be made by introducing the two wires into a little sleeve of tinned iron, fixing it to the wires by compressing it as a metal tag is fixed to a lace, and afterwards soldering; no points are then left sticking out at the ends of the joint.

Temporary joints in guttapercha-covered wire.—A piece of indiarubber tube, fitting tightly to the wire, is slid some distance up one of the wires. The ends of the wires are cleaned, twisted together, and soldered in the usual way, and the indiarubber tubing slid down so as to cover the joint.

Solder.—Equal parts of lead and tin. Never solder with acids or chloride of zinc in instruments. It is impossible to clean them away, and they finally corrode the metal. Chloride of zinc never dries completely, so that if it gets on to wood or ebonite, it spoils the insulation. All instrument work, and all jointing of covered wire or any kind of wire not freely exposed to rain, should be done with resin.

Red varnish.—For wood, interior of electro-magnet coils, galvanometers, etc., dissolve sealing-wax in alcohol at 90°; apply it with a pencil when cold in four or five coats, until the desired thickness is attained. It is better to use many coats than to make the varnish thick.

Agglomeration of wires.—A process employed in constructing coils for high resistance galvanometers. The wire is wound in layers, each one being covered with a pencil with a coating of a cold solution of gum copal in ether. It is baked, to dry it. The whole forms a sort of cake of considerable strength and high insulation resistance.

Covering of the external wires of large electro-magnets.—Large electro-magnets are generally wound with copper wire covered with a double layer of cotton. The outside layer is hardened by painting it with cold thick gum-lac varnish. It is gently roasted before a charcoal brazier. The layer thus formed is extremely hard. It is filed smooth, polished with flax and fine pumice powder, and finally varnished.

Cement for induction coils.—The proportions vary very much, but generally approximate to the following formula:

Resin	2 parts.
Wax	1 ,,

For hot countries, slightly increase the proportion of resin.

Varnish for silk.—Six parts of boiled oil and two parts of rectified essence of turpentine.

Varnish for insulating paper or tracing paper.—Dissolve one part of Canada balsam in two parts of essence of turpentine; digest in a bottle at a gentle heat, and filter before it grows cold.

Application of an insulating mixture to the coils of electrical instruments.—*Instructions issued to shops in which instruments are manufactured for the Indian Government, Telegraph Department (Schwendler).*—The empty reels are, first of all, carefully dried for five hours at a temperature of at least 230° F. (110° C.). The moment they are taken from the oven they are plunged into a melted mixture at a temperature of about 350° F. (180° C.); this mixture is composed (by weight) of:

Yellow wax	10 parts.
White wax	1 ,,

Bubbles of air appear on each reel after it is immersed. When no more air comes off, the pot is taken off the fire, and allowed to cool very slowly. A little before the mixture sets, the reels are taken out and replaced on the fire so as to allow the excess of mixture to drain away. When they look quite clean they are taken off and cooled. They are now ready to receive the wire.

When they have been wound, they are subjected to the same treatment, *i.e.* drying, immersion, and cooling. Practice has shown that the coils must undergo this treatment at least three times in order to ensure their complete penetration by the mixture.

Care must be taken during the successive heatings of the coils that the temperature does not rise too high; otherwise the mixture already present would run out instead of soaking into the coil. The drying temperature ought to be kept up to 230° F. (110° C.), and that of the insulating compound slightly diminished at each immersion. No paper should be used between the layers of wire. The evenness of the coil must be obtained by careful winding. Paper diminishes the magnetising effect of the coil, and prevents the composition from penetrating.

Summary.—It must be borne in mind that:

1. The preliminary drying of the reel and coils is necessary in order to get rid not only of moisture, but also of air, and so facilitate the penetration of the mixture. The temperature 230° F. (110° C.) answers this purpose.

2. The immediate immersion of the reels and coils in the melted mixture is necessary in order to prevent the penetration of air and moisture.

3. The slow cooling is intended to make sure that only the mixture penetrates into the coil when contraction by cooling takes place.

4. The reels and coils must remain in the mixture until no more bubbles are formed, this being the only indication showing that the mixture has filled up pores and crevices.

Oven.—The same oven is used to heat the reels, coils, and mixture. It is formed of several copper receptacles arranged in a box of the same metal. The box is filled with hot oil, which produces a uniform temperature in the receptacles in which the reels and coils are placed; each receptacle is provided with a tin grating, on which the coils rest to prevent their touching the bottom. The earthen pot containing the mixture is heated directly on the open fire.

Insulation of wires for telegraphy and telephony

(*C. Wiedemann*).—Prepare a bath of potassium plumbate by dissolving 10 gr. of litharge in a litre of water, to which 200 gr. of caustic potash have been added, and boil for about half-an-hour; it is allowed to settle, and decanted. The bath is now ready for use. The wire to be insulated is attached to the *positive* pole of a battery or electroplating dynamo, and dip a small plate of platinum attached to the *negative* pole also into the bath. The peroxide of lead is formed on the wire, and passes successively through all the colours of the spectrum. The insulation becomes perfect only when the wire assumes its last colour, which is a brownish-black.

This perfect insulation may be utilised for galvanometers or other apparatus.

Clark's composition.—For covering the sheathing of cables.

Mineral pitch	65 parts.
Silica	30 ,,
Tar	5 ,,

It is mixed with oakum in the proportion of one volume of oakum to two of composition. Its density is about 1·62. The weight in kilogrammes per nautical mile is obtained by multiplying the section in millimètres by 3.

Chatterton's compound.—For cementing together the layers of guttapercha in cable cores, an excellent insulator of fairly low inductive capacity.

Stockholm Tar	1 part.
Resin	1 ,,
Guttapercha	3 ,,

Is also used for filling up the interstices of shore-end cables. Its density is about the same as that of guttapercha, but its inductive capacity is less.

Insulating cement.—The best, according to *Harris*, is good sealing-wax.

Cement for insulators.—Sulphur, lead, or plaster of Paris, mixed with a little glue to prevent its setting too rapidly.

Muirhead's cement.—3 parts of Portland cement, 3 parts of coarse ashes, 3 parts of forge ashes, 4 parts of resin.

Black cement.—1 part coarse ashes, 1 part forge ashes, 2 parts resin.

Siemens' cement.—12 parts iron filings or rusty iron, and 100 parts sulphur.

Marine glue.—Used for battery troughs and generally as an insulating cement used at a high temperature, and, like common glue, with but little in the joint. It will stick together almost all materials, and form a strong joint. As, like pitch, it is a viscous solid and tends to flow, it should not be used thickly. One part of indiarubber is dissolved in 12 parts of benzine, and 20 parts of lac are added, the mixture being carefully heated. It may be applied with a brush. It is sold commercially in a solid form. It may either be redissolved or applied hot, using a rather high temperature and taking care not to allow it to burn.

Cement to resist heat and acids.—

Sulphur	100 parts.
Tallow	2 ,,
Resin	2 ,,

Melt the sulphur, tallow, and resin together until they are of a syrupy consistence and of a reddish-brown colour. Add sifted powdered glass until a soft easily-applied paste is produced. Heat the pieces to be joined, and use the cement very hot.

Cement used by Gaston Planté for his secondary batteries

is run *hot* on the corks and connecting strips of the secondary cells to prevent the acid from creeping.

Turner's cement	1,000 parts.
Tallow, or beeswax	100 ,,
Powdered alabaster powder	250 ,,
Lampblack (to colour it black)	2·5 ,,

Waterproofing wooden battery cells (*Sprague*).—

When the boxes are quite dry and warm, it is smeared over inside with a hot cement, composed of four parts of resin and one part of guttapercha, with a little boiled oil.

Note by translator.—The addition of boiled oil improves all substances used for this purpose which contain pitch, marine glue, or other viscous solid; tending to prevent them from flowing.

Cement for bone and ivory.—

A solution of alum, concentrated to a syrupy condition by heat. Apply hot.

Ebonite.—

Mixture of 2 to 3 parts of sulphur, with 5 parts of indiarubber, kept for some hours at a temperature of 75° C. under a pressure of 4 to 5 atmospheres. May be moulded into any desired shape. An excellent insulator, but becomes porous and spongy under the action of moisture, and loses its properties. To keep vulcanite in good order, it should be occasionally washed with a solution of ammonia, and rubbed with a rag slightly moistened with paraffin oil (*Silvanus Thompson*).

Watertight decomposition cells for electrotyping (*E. Berthoud*).—A well-made vat of oak may last for twelve or fifteen years, if it be smeared inside with the following composition:

Burgundy pitch	1,500 grammes.
Old guttapercha in small shreds . .	250 ,,
Finely powdered pumice stone . . .	750 ,,

Melt the guttapercha, and mix it well with the pumice-stone. Then add the Burgundy pitch. When the mixture is hot, smear the inside of the vat with it. Lay it on in several coats. Roughness and cracks are smoothed off with a hot soldering-iron. The heat of the iron makes the cement penetrate into the pores of the wood, and increases its adhesion. The vat will stand sulphate of copper baths, but not baths containing cyanide.

Turner's cement.—Used for fixing together pieces which have to be turned up to the same dimensions. It is thus composed: two-thirds brown resin and one-third beeswax. These proportions must be modified according to the temperature. In summer there must be less wax, and in winter less resin, so as to form a malleable cement, and tenacious enough to resist the friction of the tool on the material, also to resist the heating of the material, which is greater if it be used on a metal mandril. Wooden mandrils (of nutwood, for example) are better for this class of work. It is, however, necessary, when wooden mandrils are used, to pass the tool lightly over them, to make it smooth and of uniform thickness.

A practical test of the quality of this substance is to let a drop of the melted cement drop on to a piece of metal; when it is cold, chip it off and bend it between the finger and thumb. If it breaks, it requires the addition of more wax; if it is too plastic, of more resin. Experience alone can guide the workman to the right consistency (*Oudinet, Principes de la construction des instruments de précision*).

Composition for cushions of frictional electric machines.—Canton advises the use of an amalgam of zinc and tin. Kienmayer gives the following formula: equal parts of zinc and tin; melt, and add twice the weight of alloy of mercury. When the rubbed plate or cylinder is of vulcanite, the amalgam must be softer than when it is of glass. In France they generally use *mosaic gold* (bisulphide of tin). The amalgam must be reduced to fine powder, and applied by the aid of a little hard grease.

Cleaning, scaling, and pickling of copper and its alloys.

—A very important series of operations, in which the surface of objects which are to be electro-plated is made chemically clean, so as to ensure the adhesion of the metallic surfaces:

1. *Cleansing from grease.*—Heat the articles over a slow fire of coal-dust, baker's braise, or, better still, in an oven, up to a dull red heat. Delicate or soldered articles must be boiled in an alkaline solution of caustic potash, dissolved in ten times its weight of water.

2. *Scaling.*—The scaling bath is composed of 100 parts of water, and 5 to 20 parts of sulphuric acid at 66° Baumé. The articles may generally be dipped in the bath hot. Let them stop in the bath until they take a red-ochre colour. Articles cleaned from grease by caustic potash must be washed and rinsed with plenty of water before scaling. Henceforward the articles must not be touched with the hand. Copper hooks, or, better, glass hooks, should be used; for very small objects, stoneware or porcelain dishes.

3. *Passing through old strong water.*—This is nitric acid, weakened by former use. The articles are left in it until the red layer disappears, so that, when they have been rinsed, they only show a uniform metallic lustre. Rinse.

Passing through quick strong water.—The articles, after being well shaken and drained, are dipped in a mixture of

Nitric acid at 36° (yellow)	100 volumes.
Chloride of sodium	1 ,,
Calcined tallow (bistre)	1 ,,

The articles ought only to remain in the bath for a *few seconds*. Avoid heating, or the use of too cold a bath. Rinse in cold water.

5. *Passing through brightening or mating strong water.*—Articles which are to show a high polish are dipped for one or two seconds (shaking them) in a *cold* bath of

Nitric acid at 36°	100 volumes.
Sulphuric acid at 66°	100 ,,
Copper salt, about	1 ,,

When a mat surface is wanted, the bath is composed of

Nitric acid at 36°	200 volumes.
Sulphuric acid at 66°	100 ,,
Sea salt	1 ,,
Sulphate of zinc	1 to 5 ,,

They should be left in the bath for from 5 to 20 minutes, according to the kind of surface required. They must then be washed for some time in plenty of water. The articles present an earthy, disagreeable appearance, which disappears on dipping them rapidly in the bright bath, and then rinsing them quickly.

6. *Passing through nitrate of mercury.*—Dip the articles for one or two seconds into a bath of

Water	10 kilogrammes.
Nitrate of mercury	10 grammes.
Sulphuric acid	20 ,,

Shake before using. The bath should be richer in mercury if the articles are heavy, less rich if they are light. A badly cleaned and scaled article comes out of various colours and without metallic lustre. It is better to throw away a worked-out bath than to refresh it. After passing through the mercury bath, they must be rinsed in plenty of water, and then placed in the silvering or gilding bath.

Cleaning articles for nickel-plating (*Gaiffe*).—*Cleaning from grease.*—Rub them with a brush dipped in a thin, hot paste of whitening, water, and carbonate of soda. The cleansing from grease is perfect when the articles are easily wetted by water.

Scaling.—Copper and its alloys are scaled in a few seconds by dipping them in a bath composed (by weight) of 10 parts of water, 1 part nitric acid. For unfinished articles a stronger bath is required, composed of: water, 2 parts; nitric acid, 1 part; sulphuric acid, 1 part.

Steel, wrought and cast-iron (polished) are scaled in a bath composed of 100 parts of water and 1 part sulphuric acid. They are left in the bath until they become of an uniform grey colour. They are then rubbed with moistened pumice-stone powder, which lays the metal bare.

Unfinished steel, wrought and cast-iron must remain in the bath for three or four hours, then be rubbed with well-sifted powdered stoneware and water. The two operations are repeated until the coating of oxide has completely disappeared.

Deposition of copper on glass.—The glass is varnished with a solution of guttapercha in turpentine or naphtha, or with wax dissolved in turpentine. It is then brushed over with plumbago and put in the bath. The surface of the glass may be roughened by exposing it to the fumes of hydrofluoric acid, but this is rarely necessary.

Tempering of drills and tools *for piercing and cutting hard or tempered steel, either in the lathe or machine, when the article can only be finished after it has been tempered, such as saw-blades, etc.*—

Heat the tool to a cherry-red, then dip it in powdered resin; replace it in the fire, and repeat the operation two or three times, then throw the tool into water at a temperature of 20° C. To use the tool turn slowly, and take care to keep both tool and work well moistened with essence of turpentine. If these directions be observed a good result can easily be obtained. Take care to give but little bevel to the tool; if it is a drill, give it the shape of a conventional snake's tongue, like the ace of spades.

Black bronze.—A steel bronze can easily be put on copper by moistening it with a dilute solution of chloride of platinum, and slightly heating it. It may also be done by dipping the copper (well cleaned) into an acid solution of chloride of antimony (butter of antimony dissolved in hydrochloric acid), but the colour is sometimes violet instead of black (*Roseleur*).

Green or antique bronze.—Dissolve 30 gr. of carbonate or chloride of ammonium, 10 gr. of common salt, and the same quantity of cream of tartar and of acetate of copper in 100 gr. of acetic acid at 8° Baumé, or 200 gr. of common vinegar, and add a little water. When thoroughly mixed, the solution is daubed over the copper article which is to be bronzed, which is then allowed to dry in the open air for four-and-twenty or eight-and-forty hours. At the end of this time it will be found to be completely *verdigrised*, but in different tints. The article is then brushed all over, and especially on the parts in relief, with a waxed brush, and if necessary the parts in relief are touched with colour. The green parts, which are to be made bluer, may be lightly touched with ammonia, and those where the tint is to be deepened, with carbonate of ammonium (*Roseleur*).

Medal bronze.—The object being well cleaned, a thin paste of red oxide of iron and plumbago is applied with a pencil; it is then strongly heated; then when quite cold it is rubbed for a long time in every direction with a softish brush, which is very frequently passed over a piece of beeswax, and then over the mixture of red ochre and plumbago. This process gives a very brilliant reddish bronze, which is very effective on medals.

Bronzing iron.—The articles to be bronzed, carefully cleaned, are exposed for about five minutes to the fumes of a mixture of equal parts of hydrochloric and nitric acids. They are then heated to a temperature of 300° to 350°, until the colour of the bronze becomes visible. After cooling they are rubbed with paraffin, and again heated until the paraffin begins to decompose; this last operation is repeated six times,

If they are now again exposed to the fumes of a mixture of concentrated hydrochloric and nitric acids, tints of pale brown-red are obtained. Adding acetic acid to the other two acids, coatings of oxide of a fine yellow bronze colour are obtained. All gradations of colour from light-brown red to dark-brown red, or from light-yellow bronze to dark-yellow bronze, may be produced by varying the mixture of acids.

Professor *Oser* has covered with oxide by this process some iron rods 1·5 mètres long, and he asserts that after six months' exposure to the atmosphere of his laboratory, which is charged with acid vapours, they show no sign of being attacked (*Dingler*).

Preparation of electric-light carbons.—The problem is how to prepare carbon of higher conductivity than wood charcoal, and which, if not absolutely free from hydrogen, is at least free from all mineral substances. To attain this end three methods may be employed. 1st. The action of dry chlorine on carbon at a white heat. 2nd. The action of fused caustic soda. 3rd. The action of hydrofluoric acid in the cold on carbon cut into pencils immersed in it for some time. The use of chlorine answers perfectly for finely-divided carbon. By the double influence of chlorine and a high temperature the silica, alumina, manganese, alkaline oxides, and metallic oxides are reduced and transformed into volatile chlorides, and the hydrogen remaining in the carbon is transformed into hydrochloric acid, which is carried off with the chlorides.

M. Jacquelain applies this method to solid carbon by directing a stream of dry chlorine for at least 30 hours on to a few kilogrammes of retort carbon cut into prismatic pencils, and heated to a white heat.

This first operation leaves the carbon full of cavities, which have to be filled up, in order to restore as far as possible the original conductivity and feeble combustibility of the carbons; this is attained by submitting them, after the chlorine purifying process, to the action of a hydrocarbon which circulates slowly in the form of vapour for five or six hours over the pencils heated to a white red heat in a cylinder of refractory clay. The vaporisation of the hydrocarbon (heavy coal oil) must go on slowly, so that the decomposition may go on at the highest temperature, and so as to produce but a small deposit of carbon, otherwise all the pencils would be covered with a layer of hard carbon, thick enough to fix them all together into a solid block, and thus render them useless. The action of caustic soda with three equivalents of water fixed in vessels of sheet or cast-iron is more rapid, converting silica and alumina into alkaline silicates and aluminates; by repeated washing with hot distilled water, the alkali, which has soaked in, is removed together with the silicates and aluminates; then by washing with very

SOLUTION FOR CHEMICAL TELEGRAPHS. 309

weak hot hydrochloric acid and water all the iron oxide and earthy bases are removed; and then a few washings with hot distilled water remove the remaining hydrochloric acid.

The operation of purifying retort carbon by hydrofluoric acid is most simple. The carbon pencils are immersed in hydrofluoric acid, diluted with twice its weight of water, and left for twenty-four to forty-eight hours, at a temperature of 15° to 25° C., in a rectangular covered leaden vessel. They are then washed in plenty of water, and then with distilled water, dried and carbonated for from three to four hours, if the earthy substances removed by the hydrofluoric acid are not in very great quantity. But the use of this acid even when diluted with twice its weight of water requires great care and precaution.

Solution for paper for chemical telegraphs.—One part saturated solution of ferrocyanide of potassium, one part saturated solution of nitrate of ammonium, two parts water.

Translation of the Morse character into letters (*Commandant Perein*).—The ingenious diagram below enables the letter corresponding to a Morse sign to be rapidly found. The diagram is thus used:

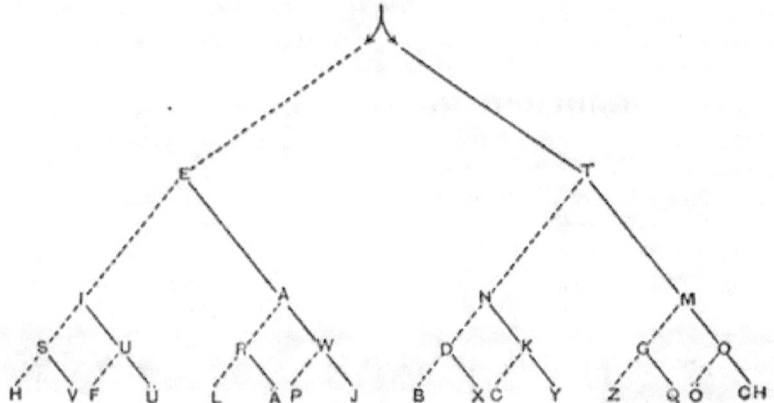

Fig. 48.—Diagram for translating the Morse Alphabet.

In order to find what letter corresponds to a given sign, starting from the top of the diagram, each line is traced down to a bifurcation, taking the right hand line of each bifurcation for a *dash*, and the left hand line for a *dot*, and stopping when the dots and dashes are used up. Thus, for example, the signal — - - leads us to the letter *d*, the signal - — — — to the letter *j*, and so on.

Fixing electric bell-wires in houses.—Copper wire $\frac{1}{1\frac{1}{2}}$th of a millimètre in diameter is generally used for the wires from the battery, and wires of $\frac{9}{10}$th or 1 millimètre in diameter for the branches to the pushes. Insulated wire should always be used. One of the best bell-wires is covered with indiarubber, over which is a layer of braided cotton soaked in paraffin; guttapercha wire is very good, but simple cotton-covered wire is usually all that is wanted; nothing, indeed, is better in dry places, if it be given a coat of shellac varnish after it is put up. The wires should always be kept an inch or two apart; it is unadvisable, in passing through walls, door frames, etc., to put both wires through the same hole, even if guttapercha wire be used. The wires may be fixed with small staples, but care must be taken that these are not hammered in so tightly as to cut through the insulation. Wherever joints have to be made, resin alone should be used for soldering, and the joint be covered with gutta-percha tissue or several coats of shellac varnish.

Bells.—The best are mounted on bed plates of metal, which avoids the disarrangement of their adjustment, due to the play of wooden frames. They should be fitted with a set screw to prevent the contact screw from getting out of adjustment. They should never be fixed directly to a damp wall; whenever the wall is damp, a slab of wood painted with oil colour should be interposed.

Static induction machines.—These machines give a continuous current, like batteries and magneto and dynamo machines; the current is very small, but the electromotive force is very high. Thus the ordinary laboratory type of Holtz machine has a constant e. m. f. of about 50,000 volts at all speeds, but the current strength increases in proportion to the speed of rotation; at 120 revolutions per minute its internal resistance is 2,180 megohms, at 450 revolutions it falls to 646 megohms. According to Kohlrausch's experiments the maximum current furnished by a Holtz machine can only decompose ·0035 microgramme of water per second, which corresponds to a current strength of about 40 microampères.

Ink for writing on glass.—Dissolve at a gentle heat 5 parts of copal in powder in 32 parts of essence of lavender, and colour it with lampblack, indigo, or vermilion.

Ink for engraving on glass.—Saturate commercial hydrofluoric acid with ammonia, add an equal volume of hydrofluoric acid, and thicken with a little sulphate of barium in fine powder. A metal pen may be used; the ink bites almost instantaneously. The glass then only requires to be washed in water.

Spray producer.—A common spray producer, which may be bought at any chemist's for a few pence, is very handy for fixing magnetic figures on gummed paper or glass.

Coppering by simple immersion.—The following process is often used in order to protect iron and steel articles from rust, and give them the appearance of copper, when it is not required to obtain a lasting deposit or perfect adhesion.

Prepare the articles by brushing them hard with petroleum, and wiping them in hot sawdust, then dip them for one minute only into a saturated solution of sulphate of copper, to which half its volume of acidulated water has been added. Take them out and wash them quickly by dipping them in boiling water and wiping them with hot sawdust. Very small articles can often be coppered by rubbing them well in sawdust well moistened with an acidulated solution of sulphate of copper.

Another process, which is especially suited for cast-iron articles, consists in using a solution of 10 parts nitric acid, 10 parts chloride of copper, and 33 parts of hydrochloric acid. The articles are dipped several times, and wiped after each immersion with a woollen rag. When iron wire is thus coated it should be afterwards re-drawn so as to consolidate the layer of copper, and make it adhere better.

To coat large articles such as statues, candelabra, etc., the following solution is used:

Water	25 litres.
Potassic tartrate of soda	8 kilog.
Caustic soda	3 ,,
Sulphate of copper	1·25 ,,

The mixture is carefully stirred, and time given for the complete solution of the ingredients; the articles should be dipped in the solution by zinc wires. The work goes on slowly. Five hours are required for a uniform deposit.

After being taken out of the bath the articles are carefully washed and dried (*H. Fontaine*).

BIBLIOGRAPHY

OF THE PRINCIPAL WORKS CONSULTED BY M. HOSPITALIER.

PERIODICALS.—*Annales télégraphiques.*—*La lumière électrique.*—*L'Electricien.*—*L'Electricité.*—*Bulletin de la Société française de physique.*—*Journal de physique.*—*Annales de physique et de chimie.*—*Comptes rendus de l'Académie des sciences.*—*Revue scientifique.*—*Revue industrielle.*—*La Nature.*—*Le Génie civil.*

Journal télégraphique.—*Archives des sciences physiques et naturelles de Genève.*

Telegraphic Journal and Electrical Review.—*Journal of the Society of Telegraph Engineers and Electricians.*—*The Electrician.*—*Philosophical Magazine.*—*Proceedings of the Royal Society.*—*Nature.*—*Engineering.*—*The Engineer.*

Elektrotechnische Zeitschrift.—*Centralblatt für Elektrotechnik.*

GENERAL TREATISES.—*Traités de physique,* by Jamin, Daguin, Angot, Ganot.—*Traité d'électricité,* by De La Rive.—*Traité d'électricité et de magnétisme,* by A.-C. Becquerel.—*Traité d'électricité et de magnétisme,* by J. G. H. Gordon (French translation by M. J. Raynaud).—*Traité d'électricité statique,* by M. Mascart.—*Traité d'électricité et magnétisme,* by MM. Mascart and Joubert.—*Traité pratique d'électricité,* by C. M. Gariel.—*Cours d'électricité,* by E. Duter.—*Recherches sur l'électricité,* by M. G. Planté.—*Electricity and Magnetism,* by J. Clerk-Maxwell.—*Electricity and Magnetism,* by Fleeming Jenkin.—*Elementary Lessons in Electricity and Magnetism,* by Silvanus Thompson.—*The Student's Text-book of Electricity,* by Noad and Preece.—*Electricity,* by Dr. Fergusson.—*Magnetism and Electricity,* by Dr. Guthrie.

UNITS.—*Grandeurs électriques,* by E. Blavier.—*Units and Physical Constants,* by Everett.—*Exposé sommaire de la mesure électrique en unités absolues,* by J. Raynaud.—*Sur la mesure pratique des grandeurs électriques,* by W. H. Preece.—*Les mesures électriques,* by T. Rothen.—*Reports of the Committee on Electrical Standards of the British Association,* edited by Fleeming Jenkin.

INSTRUMENTS AND METHODS OF MEASUREMENT.—*Manuel d'électrométrie industrielle*, by R. V. Picou.—*Electric Testing*, by H. R. Kempe.—*Testing Instructions*, by Schwendler.—*Epreuves électriques des câbles télégraphiques*, by W. Hoskiær.

APPLICATIONS.—*Exposé des applications de l'électricité*, by Th. du Moncel.—*Les principales applications de l'électricité*, by E. Hospitalier.—*L'électricité et ses applications*, by H. de Parville.—*Electricity: its Theory, Sources and Applications*, by John T. Sprague.—*Traité de la pile électrique*, by A. Niaudet.—*Traité théorique et pratique des piles électriques*, by A. Cazin and A. Angot.—*Guide pratique du doreur, de l'argenteur et du galvanoplaste*, by A. Roseleur.—*Electro-Metallurgy*, by G. Gore.—*Electroplating*, by Urquhart.—*Machines électriques*, by A. Niaudet.—*Die Magnet und dynamo-elektrischen Maschinen*, by Dr. H. Schellen.—*L'électricité comme force motrice*, by Th. du Moncel and Frank-Géraldy.—*Electric Transmission of Power*, by Paget Higgs.—*Eclairage à l'électricité*, by H. Fontaine.—*L'éclairage électrique*, by M. Th. du Moncel.—*La lumière électrique*, by Alglave and Boulard.—*Electric Illumination*, by J. Dredge.—*Electric Lighting*, by Sawyer, Schoolbred.—*Traités de télégraphie électrique*, by MM. Mercadier, Blavier, Th. du Moncel, Breguet, Gavarret.—*Le siphon-recorder*, by Ternant.—*Systèmes télégraphiques*, by Ch. Bontemps.—*Electricity and the Electric Telegraph*, by G. B. Prescott.—*Manuel de télégraphie pratique*, by R. S. Culley.

ANNUALS, POCKET-BOOKS, AND TABLES.—*Electrical Tables and Formulæ*, by Latimer Clark and Robert Sabine.—*Annuaires du bureau des longitudes et de l'observatoire de Montsouris.*—*Agendas du chimiste, des postes et télégraphes*, d'Oppermann.—*Carnet de l'officier de marine.*—*Rules and Tables*, by Rankine.—*Formulaire de l'ingénieur*, by Ch. Armengand.

VARIOUS.—*Comptes rendus des travaux du Congrès international des électriciens.*—*Procès-verbaux de la réunion internationale des électriciens.*—*Les phares électriques*, by E. Allard.—*Note sur les appareils photoélectriques employés par les marines militaires*, by MM. Sautter and Lemonnier, etc., etc.

INDEX TO TABLES.

Acceleration due to gravity and the length of the seconds pendulum, Values of, 36.
Area, Units of, 35.
Ayrton and Perry's motors, 251.

B.

Barometer, Mean height of, 155.
Batteries, Constants and work of some known forms, 203.
——, E.m.f. of amalgams of potassium of zinc, 200.
——, —— of Grove's cell, 200.
——, —— of some two-fluid cells, 202.
——, —— of various, 201.
——, Porous pot, 192.
——, Testing, 206.
Birmingham wire gauge, 172.
Bronze-phosphor and silicium, Properties of, 177.
Brush machine, 242.
Burgin machine, 242

C.

Candles, electric, Tests of, 268.
Capacities, Specific inductive, 179.
Capacity, Units of, 34.
Carbon (Cylindrical), resistance per metre, 165.
Carbons, Experiments with bare and plated, 264.
Chemical and electro-chemical equivalents, 212.

Coefficients of expansion of some solids, 157.
Conductivity (relative) of copper, 163.
—— of metals, 164.
—— of solutions, 166.
Conductors, Diameter and resistance of some, 178.
——, Loss of energy in, 227.
Copper, Conductivity (relative) of, 163.
——, Resistance of pure, at 0° C., 174.
—— sulphate, Densities of solutions of, 155.
—— ——, Specific resistance of, 167.
Current strength, Units of, 46.

D.

Densities of solutions of common salt, 154.
—— —— of copper sulphate, 155.
—— —— of nitric acid, 153.
—— —— of zinc sulphate, 153.
Diamagnetic substances, 180.
Diameter and resistance of some conductors, 178.
Duprez's experiments, 256-7.
Dynamo-electric machines:
  Brush, 242.
  Burgin, 242.
  Edison, 240.
  Edison-Hopkinson, 241.
  Elphinstone-Vincent, 242.
  Ferranti-Thomson, 243.
  Gramme, 178.

Dynamo-electric machines (contd.):
  Gulcher, 237.
  Heinrichs, 237.
  Méritens, 244.
  Schuckert, 238.
  Siemens, 237, 240, 243.
Dynamo-electric machines, Size of wire for, 235.
—— machines. (See also Machines.)

E.

Ed'son-Hopkinson machine, 241.
Edison's lamp, 271.
—— machine, 240.
Electrical resistance, Units of, 45.
Electric candles, Tests of, 268.
—— -light carbon, Resistance of, 165.
Electrification, Influence of length of time of, 170.
Electro-chemical equivalents, 212.
—— -magnetic units, 41.
E.m.f. of amalgams of potassium and zinc, 200.
—— of some two-fluid cells, 202.
—— of various cells containing only one electrolyte, 201.
Electrotyping: coppering baths, 218.
——, weight of deposit per hour and strength of current, 217, 221.
Elphinstone-Vincent machine, 242.
Energy, heat, and work, Units of, 51.
——, Loss of, in conductors, 227.

F.

Ferranti-Thomson machine, 243.
Force and weight, Units of, 38.
French and English units of length, 33.
—— —— units of volume and capacity, 34.

G.

Gases and vapours, Specific gravity of, 154.

Geometrical formulæ, 137.
Gramme and Siemens motors, 251.
—— machine, 236.
—— —— and projectors used in the French navy, 267.
Gravities, Specific, 148-152.
Grove's cell, E.m.f. of, 200.
Gulcher machine, 237.

H.

Heat disengaged by the combination of gramme with chlorine, 159.
—— —— by the oxydation of gramme, 159.
——, work, and energy, Units of, 51.
Heinrichs machine, 237.

J.

Jauge carcasse, 173.

L.

Lamps, incandescent Edison, Tests of, 271-2.
—— ——, Tests of, 272.
—— ——, Tests of Siemens and Halske's, 272.
—— (See Edison, etc.)
Length, Units of, 33.
Logarithms, 135-6.

M.

Machines and lamps, Tests of continuous current, 266.
—— (See Dynamo-electric machines.)
Magnetic substances, 180.
Melting and boiling points of common substances, 158.
Méritens machine, 244.
Metals, Conductivity of, 164.
Motors, Ayrton and Perry's, 251.
—— Gramme and Siemens, 251.

INDEX TO TABLES. 317

N.

Nitric acid, Densities of solutions of, 153.
———— ————, Resistance of, 167.
Numbers, their reciprocals, etc., 131.

P.

Phosphor-bronze, Properties of, 177.
Polygons, Radii and areas of regular inscribed, 138.
Pressure, Units of, 50.

R.

Radii and areas of regular inscribed polygons, 138.
Relay, Schwendler's experiments on a Siemens polarised, 285.
Resistance: List of common bodies in order of decreasing conductivity, 161.
—— of common metals and alloys, 162.
—— of cylindrical carbons per mètre, 165.
—— of nitric acid, 167.
—— of solutions of copper sulphate, 167.
———— —— of sulphuric acid, 167.
———— —— of zinc sulphate, 168.
—— of wires of pure annealed copper at 0° C., 174.
——, Specific, of water and ice, 168.
——, Strength of received currents and equivalent insulation resistances, 278.
——, Units of, 45.
—— (See also Conductors.)

S.

Salt, common, Densities of solutions of, 154.
Schuckert machine, 238.

Siemens machines, 239, 240, 243.
———— and Halske's incandescent lamps, Tests of, 272.
Silicium bronze, 177.
Sines and tangents, 139.
Solders, 294.
Solids, Cubic coefficients of expansion of, 157.
Solutions, Conductivity of, 166.
Specific gravities, 148-152.
———— gravity of gases and vapours, 154.
———— inductive capacities, 179.
Submarine cables, Details of some, 282.
Subterranean cables with seven conductors, 280.
Sulphuric acid, Specific resistance of solutions of, 167.

T.

Temperature, 158.
——, Measurement of, 156.
—— (See Melting and boiling points of common substances.)
Temperatures, Determination of high, 157.
Testing batteries, 203.
Tests of continuous current machines and lamps, 266.
Thermo-electric powers for calculating, 225.
Thermometer, 49.
—— scales, Fahrenheit and Centigrade, 156.
Transmission of power, 254.
———— ————, Duprez's experiments, 256-7.
Triangles, Solution of, 142.
Trigonometrical formulæ, 140.

U.

Units, Electro-magnetic, 41.
—— of area, 35.
—— of current strength, 46.
—— of electrical resistance, 45.
—— of energy, heat, and work, 51.

Units of length, 33.
—— of pressure, 50.
—— of volume and capacity, 34.
—— of weight and force, 38.
—— of work, 39.

V.

Values of the acceleration due to gravity and the length of the seconds pendulum, 36.
Volume and capacity, Units of, 34.

W.

Water and ice, Approximate specific resistances of, 168.

Weight and force, Units of, 38.
Wire-gauge, Birmingham, 172.
—— jauge carcasse, 173.
——, Size of, for dynamo-electric machines, 235.
Work, heat, and energy, Units of, 51.
——, Units of, 39.

Z.

Zinc sulphate, Densities of solutions of, 153.
—— ——, Specific resistance of solutions of, 168.

www.ingramcontent.com/pod-product-compliance
Lightning Source LLC
Chambersburg PA
CBHW021159230426
43667CB00006B/468